Digital Economic Dynamics

Paul J. J. Welfens · Mathias Weske
Editors

Digital Economic Dynamics

Innovations, Networks and Regulations

With 29 Figures and 16 Tables

Springer

Professor Dr. Paul J. J. Welfens
University of Wuppertal
EIIW –European Institute
for International Economic Relations
Rainer-Gruenter-Straße 21
42119 Wuppertal
Germany
welfens@eiiw.uni-wuppertal.de

Professor Dr. Mathias Weske
University of Potsdam
HPI – Hasso Plattner Institute
Prof.-Dr. Helmert-Straße 2–3
14482 Potsdam
Germany
mathias.weske@hpi.uni-potsdam.de

Cataloging-in-Publication Data
Library of Congress Control Number: 2006930880

ISBN-10 3-540-36029-8 Springer Berlin Heidelberg New York
ISBN-13 978-3-540-36029-2 Springer Berlin Heidelberg New York

Springer is a part of Springer Science+Business Media

springer.com

Printed in Germany

Hardcover-Design: Erich Kirchner, Heidelberg

SPIN 11790969 43/3153-5 4 3 2 1 0 – Printed on acid-free paper

Contents

Introduction 1
Paul J.J. Welfens and Mathias Weske

Software Development Dynamics: Current Trends and Future Developments 5
Dominik Kuropka and Mathias Weske

Digital Social Market Economy – Towards a New Economic System 25
Paul J.J. Welfens

Market Analyses under the New European Communications Framework - Some Conceptual Issues 63
Wolfgang Briglauer

Liberalisation of Telecommunications and Innovation Dynamics in Denmark 91
Morten Falch and Anders Henten

Regulations of Telecommunications and Market Dynamics in Germany 119
Friedhelm Dommermuth

Innovation in the European and Japanese Mobile Telecommunications Industry 129
Richard Tee

Collective Dominance, Market Analysis and the 2002 EU Framework Directive: The Case of Mobile Access and Call Origination in Ireland 141
Chris Doyle

Credible Threats as an Instrument of Regulation for Network Industries 171
Justus Haucap, Ulrich Heimeshoff and André Uhde

Information Society 2015: Towards Increased Participation 203
Jaroslaw K. Ponder

List of Contributors 211

Introduction

Information and communication technology (ICT) as well as the significance of a growing digital communication network have generated enormous economic impulses. While hardware investment is important in this respect, there is an even more crucial role of software; organizing value-added based on software requires standardization and modularization, thus causing rationalization effects as a side-impact thereof. Software solutions are more flexible than hardware-based problem solving. Falling relative prices of hardware – reflecting high technological progress – and the benefits from global network effects in the software industry (which also benefits from massive scale effects) can be expected to serve as the major driving forces of productivity increases and economic growth in the 21st century. While software markets have specific problems from the perspective of competition policy, the telecommunications markets also pose special problems. Traditionally viewed as natural monopolies, the view of economist and policy makers changed in the late 20th century. In fixed network telecommunications, there is not a serious problem with competition provided there is full competition from cable TV networks as is the case in the US, the Netherlands, and – to a lesser extent – in the UK. Under the pressure of the European Commission, EU countries opened up fixed network telecommunications markets in 1998. The UK had already started liberalization in 1984 as did the US. The challenge for EU countries is to achieved sustained competition. Regulation serves several goals, and depending on the regulatory policy chosen there could be positive effects or negative effects on Schumpeterian dynamics. The dynamic perspective of electronic communications regulations should not be neglected over the traditional emphasis on static efficiency aspects.

Competition in telecommunications requires to some extent ex ante intervention of competition authorities. Asymmetric regulation of fixed line telecommunications is assumed to bring about static and dynamic efficiency; newcomers compete mainly through prices while the incumbent operator enjoys a dominant market position in an initial situation for which it additionally has leverage through vertical integration. Regulation typically concerns interconnection rules and access to essential facilities. The latter refers to the case of fixed line telecommunications for which there are particular problems in the local loop where the incumbent operator typically has a market share of 80-90% in EU countries. Thus further regulatory rules typically refer to resale. It is true that the increasing relevance of mobile telecommunications has reinforced competition to some extent, but mobile telecommunications also has problems as has become evident in the context of very high international roaming charges in the EU. Fixed-mobile substitution is only relevant for part of the telecommunications market. With broadband services delivered alternatively through DSL fixed-line networks or through cable TV networks, there are new prospects for digital innovation, networking and competition; digitization has blurred market demarcations. Voice telephony, data traffic and TV have finally become a single market, and this implies enormous challenges in a modern market economy. Moreover, there are also new opportunities for digital trade and foreign investment. At the bottom line, the telecommunications market – strongly linked with the internet market – has become one of the most dynamic markets in the early 21st century.

Sustained competition in telecommunications broadly defined is important for innovation, diffusion and growth in OECD countries and worldwide. Broadband competition – most relevant for fast internet and in turn crucial for cheap VOIP telephony – cannot be achieved without adequate regulation of electronic communications which includes both fixed line (DSL) telephony and cable TV networks.

Paul J.J. Welfens takes a look at some key topics of a digital market economy. He considers the role of structural change and growth as well as specifics of the digital economy, network effects, software dynamics and innovation aspects. He takes a critical view of the digital media society which is partly causing negative external effects. Finally, he emphasizes key policy issues including major competition policy aspects which the authors presented at an FCC seminar in Washington, DC. It is argued that differentiated two-part tariffs are welfare-maximizing and a better policy approach than complex Ramsey prices.

Dominik Kuropka and Mathias Weske discuss the increasingly important role of software technology in the context of the dynamic development and flexible operation of software systems, which are instrumental for the success of modern business organizations. Software engineering is described as the key discipline for structured analysis, development and deployment of software systems. In particular, a sample software methodology is shown that consists of a set of interrelated phases that are being conducted during the development of a software system. Service-oriented computing is introduced as one of the major recent developments in software technology. One of the key expectations is that service-oriented software systems can be developed and maintained with reduced cost as well as enhanced flexibility and dynamism, thus supporting two major goals of modern software development. The comprehension of processes and their properties has been identified as an additional aspect of complex software systems. The authors briefly discuss the combination of service-orientation and process-orientated software systems design and development. This contribution closes with a discussion of a set of research-stage topics that are related to the vision of a service-oriented market including automated on-demand integration and invocation of services as well as automated service composition. These topics are highly relevant for dynamic software development in the years to come.

Wolfgang Briglauer's analysis directs focus to EU framework regulation. Under the new European regulatory communications framework which came into effect in the summer of 2003, market analyses are subject to a different and much wider analytical scope. This article discusses conceptual aspects and practical experiences which arose after the first series of market analyses conducted under the new regulatory framework. According to the three stage analysis procedure of the European Commission, selected issues are discussed sequentially.

Richard Tee compares the development of mobile services in Japan with the contrasting paths witnessed in European markets. Using a strategic networks perspetive, he focuses on the role of the network operator and its central role in the

Japanese innovation system. By contrast, the European innovation system can be characterized by protocols and standards, largely stemming from the mobile handset manufacturers. The author has two main goals. First, he attempts to extend the existing work on mobile (data) services by critically examining various arguments commonly proposed for explaining the opposing Japanese and European mobile internet state of play. Second, he gives a characterization of both market structures and briefly looks at various policy measures that might be taken.

Morten Falch and Anders Henten discuss the problems of liberalization and innovation dynamics in Denmark and the Nordic countries, respectively. In Europe, Denmark is known for being quite innovative and open for dynamic domestic and foreign investors. The authors show how regulations implemented have affected innovations in telecommunications services markets, emphasizing both process and product innovations. The description and analysis of the Danish and Nordic telecommunications markets is quite interesting, as it shows how important broadband innovation dynamics are and that regulation as well as competition are quite relevant in this context. Moreover, their contribution demonstrates that adequate incentives for rising R&D expenditures are crucial in the digital economy with changes in market structures affecting the division of labor between network operators and equipment makers.

Germany has been one of the most dynamic telecommunications markets in Europe since 1998. Friedhelm Dommermuth focuses on several points in this respect. Two aspects are discussed in order to illustrate telecommunications regulation and the continued development of markets in Germany. On the one hand, essential developments in the telecommunications sector will be described retrospectively. On the other hand, elements of the Telecommunications Act (TKG) effective since the end of June 2004 will be referred to as examples likely to be of central importance to the future development of the telecommunications sector, namely the consultation and consolidation procedure in addition to the consistency requirement.

The contribution by Christopher Doyle presents a discussion of the national regulatory authority's market review of the mobile access and call origination market in Ireland over the course of 2004. The market review was one of the first to result in a notification of joint dominance to the European Commission (EC). The EC Article 7 Task Force accepted ComReg's reasoning, and the Commission did not raise any serious doubts about the decision. However, the affected mobile network operators designated as having Significant Market Power (SMP) (O_2 and Vodafone) appealed the decision as did another operator, Meteor, who was also designated as having SMP. Doyle examines ComReg's assessment of the structural characteristics of the market in detail, concluding with the argument that the hurdle in case law for finding joint dominance had been set at a high level, thereby presenting regulators such as ComReg with the significant challenge of finding joint dominance within the context of ex ante regulation.

Justus Haucap, Ulrich Heimeshoff and André Uhde discuss the role of credible threats as an instrument of regulation for network industries. The authors take into account institutional economics and provide a broad discussion of many key regulatory issues. By comparing important developments in New Zealand and Germany – and taking into account both telecommunications and electricity –, the authors draw on rich policy experiences. Thus the reader gets a better understanding for the cost and benefits of heavy-handed regulation compared to a light regulatory approach. For the European and global regulatory development, this discussion is of considerable importance. With technology dynamics expected to continue at a high pace and digital markets changing over time, one may anticipate an ongoing international regulatory discussion.

Jaroslaw Ponder from the International Telecommunications Union has highlighted some of the key aspects of the World Summit on the Information Society. The two-stage approach of the UN of avoiding an increasing North-South digital divide is quite crucial for a networked world economy offering equal opportunities across countries.

This book is largely the result of a Wuppertal workshop co-financed by Jean Monnet funds from the European Commission. Support by the European Institute for International Economic Relations (EIIW) at the University of Wuppertal and the Hasso Plattner Institute at the University of Potsdam is also appreciated. One should emphasize that interdisciplinary research visible in this publication is urgently needed if we are to better understand the dynamics of the digital world economy and of high technology dynamics. Economic and technological dynamics should be considered in an interdependent perspective.

We greatly appreciate the excellent editorial assistance provided by Michael Agner and Julia Tillmann. Furthermore, Michael Vogelsang (EIIW) has been an excellent project manager.

Building new digital intellectual bridges is a wonderful challenge which has been a consistent task of the EIIW over more than a decade. Progress in the digital world promises to bring about enormous economic, social and institutional challenges which should be anticipated both by the business community and policymakers. International aspects will gain importance in the long run, and the expansion of internet-based telecommunications will reinforce long term needs for global regulatory cooperation; the latter does not rule out institutional competition.

Wuppertal/Washington DC and Potsdam, June 2006

Paul J.J. Welfens and Mathias Weske

Software Development Dynamics: Current Trends and Future Developments

Dominik Kuropka and Mathias Weske[1]

1 Introduction 5
2 Software Engineering 6
3 Service-oriented Computing 9
4 Process-oriented Information Systems 14
5 The Future: Semantics, Ontology and Planning 17
6 Conclusions 21
References 22

1 Introduction

In a digital networked economy companies rely increasingly on software systems to collaborate with partner companies, to provide services to their customers and overall to achieve their business goals. While this trend has been around for years, it continues to have implications on the way software systems are designed and developed, as well as on the business models companies are based on. On the other hand, software technology provides new opportunities for companies and new methods of doing business. This paper investigates the relationship between recent achievements in software, information systems, and process technology on the one hand and the new opportunities facilitated by these technologies on the other. To provide one step forward towards understanding this relationship, this paper discusses recent trends and technologies in software, information systems and process technology. Rather than focusing on the technical details of these domains, the fundamental concepts and the new opportunities are discussed.

The remainder of this paper is organized as follows: In Section 2 software engineering methods and techniques are introduced by comparing software engineering to traditional engineering and by presenting a basic software methodology. Section 3 is devoted to service-oriented computing, a recent trend to facilitate convenient integration of services provided by heterogeneous software systems of various organizations. Today service computing is regarded as one of the main driving forces of digital modernization. While service computing provides rather technical solutions, the business value to a large extent stems from the business processes that are operated by an organization. The management of business processes is a challenging endeavour that involves multiple levels of abstraction, from business issues to technological issues. An overview of process-oriented informa-

[1] The authors are grateful to Hilmar Schuschel for contributing the subsection on automated process planning.

tion systems is presented in Section 4. The added value of the approaches presented in Sections 3 and 4, (i.e., service computing and process computing), comes from an integration of these. Recent trends in doing this make use of areas such philosophy and linguistics.

2 Software Engineering

During the past decades increasingly large and complex software systems were developed to fulfil the requirements of an increasingly dynamic market. Today, most companies in their daily business to a large extent rely on the function, availability and flexibility of their software systems. This trend pinpoints the need for a systematic approach to software development. Such approaches are summarized by the term Software Engineering. While there are definitions for this term focusing on different aspects of software development, there is agreement that software engineering is "the systematic approach to the development, operation, maintenance and requirement of software" (ANSI/IEEE, 1983).

Whereas software engineering is similar to other engineering disciplines such as civil, apparatus or construction engineering, there are some significant differences to those classical engineering domains. Probably the most challenging of these is the fact that the resulting product of software engineering is very complex and not tangible, as opposed to products of classical engineering disciplines. For instance a bridge or an automobile are also complex but are, in fact, tangible. In addition, many people with different expertise and competence collaborate to design and develop an engineering product. The fact that software products are intangible makes the communication between the involved people quite hard, because the involved parties have to communicate about non-physical things. These things can only be represented by some kind of a metaphorical and abstract language, typically a modelling language. This situation is made worse by the fact that there are at least three types of partners involved in the software engineering process, each with different view points and backgrounds. These partners are the client, developer and user (SCHACH, 1993).

- The *client* is the individual or the organization, who wants a software product to be developed and who provides the requirements for the definitions of the product. The problem for the client is that he typically has no deep understanding of software technology or programming. So he is able to specify the requirements on a high application domain specific level, but he is usually not able to explain how the requirements can be achieved.
- The *developer* is responsible for the whole software engineering process and especially for the programming of the software system. They are usually lacking deep application domain specific knowledge and practical experiences, but they have an extensive knowledge on software engineering, computer systems and programming.

- The *user* has to work with the deployed system. In general, users may be very heterogeneous regarding their experiences in computer usage in general and application domain knowledge in particular.

The success of software engineering highly depends on successful communication between those three stakeholders. For this reason the identification of different phases, the arrangement of the involved parties and the expected documents describing what has to be delivered are crucial aspects in software engineering. According to (SCHACH, 1993), eight phases can be identified:

1. *Requirements Phase*: During this phase the client, assisted by the designers and developers has to specify the requirements for the product in a dedicated requirements document.
2. *Specification Phase*: When designers and developers feel that they comprehend the client's needs, a specification document is delivered. This document explicitly describes the functionality of the product and lists any constraints that must be satisfied.
3. *Planning Phase*: After the functionality of the product is more precisely defined, the project (including the estimated cost and time needed for producing the software) has to be planned. Prior to this stage, a realistic planning and calculation of return on investment, which is of high importance for the client, is not feasible.
4. *Design Phase*: While the specification of a product describes what the product is to do, the aim of the design phase is to determine how the product will do it. During this phase the internal structure and architecture, the algorithms and data structures within the software product have to be chosen by the designers.
5. *Implementation Phase*: During this phase the developers will implement the system and run tests on a small scale to ensure proper functionality of modules.
6. *Integration and Testing Phase*: The next stage is up to the developers to combine the modules and to determine if the product as whole runs correctly.
7. *Maintenance / Operation Phase*: At the beginning of this phase the developers deploy (or rollout) the product and the users start working with the system. To ensure proper working, a small group of developers have to keep an eye on the system and make some smaller modifications if needed. User training is typically done before rollout.
8. *Retirement*: After years of service the point comes where further maintenance of the software product will be not cost-effective any more. This can be caused through changes in the environment or business-model, new technologies or lose of control over the system through bad documentation or change management. It's the task of the developers to create possible scenarios for a migration of data and functionality from the old system to a new one before the old system is shut down.

One important part of a software engineering methodology is the content and organization of the phases. A well-known model is the *waterfall model* (ROYC, 1970). The main idea of this model is that phases are executed in a sequential order. A phase can only be finished if the activity of this phase has been entirely settled and a final document for this phase has been delivered. In case of mistakes in previous phases the active phase has to be aborted and we have to move back to solve the problem in the directly preceding phase. (Refer to figure 1) After the problem has been solved, the previously aborted phase can be continued and adapted to the new results. This whole procedure may cascade if the solution to a problem forces changes in a preceding phase which is more than one step before the actual phase.

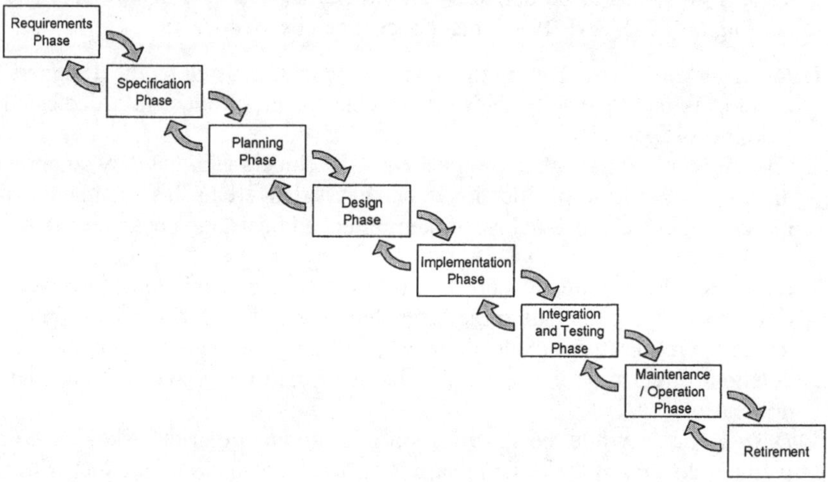

Fig. 1. Waterfall model

The benefits of this model are its simple structure and the low management efforts in conducting it. Further the model is document centric, which facilitates good documentation of the produced software system. Negative aspects of the waterfall model are the long period of time until the first parts of the system are implemented and presentable to the client. This also implies that the system fundament is set up only from theoretical concepts, which may lead to problems during the implementation, especially if new or insufficiently structured problems are addressed. It is a fact that mistakes made in early software engineering phases can be cheaply correctly in those early phases but the costs rise rapidly when the mistakes have to be corrected in later phases.

For this reason a number of other methodologies have been developed, which are able to provide practical experiences and tests in earlier stages. The key innovation of those models in comparison to the waterfall model is the use of prototypes. The *rapid prototype* software engineering process model schedules the creation of a prototype at the beginning of the software engineering process to ease the collection of requirements and the specification of the system. A huge benefit of prototyping in general is that a presentable result is created at early stages. Such a prototype can be used to gain a consensus with the client regarding the requirements of the system that is more useful than just using textual or graphical representations of the system. This is justified by the fact that it is easier for a client to identify a proper system when seeing some samples of the system, than explain from scratch what requirements have to be satisfied. Other process models such as the *incremental model* or the *spiral model* (BOEHM, 1988) use prototypes extensively in nearly every phase to reduce the development risk. For example, this can be achieved by ensuring (by the use of prototypes) that theoretical specifications are adequate and can be implemented.

Today, prototyping is a widely spread approach in the software industry, especially in creating standard software like office tools, but also, for instance, database applications. The software products in this branch seem "never to be finished", in fact it is sometimes sensible to claim that vendors are selling polished prototypes instead of finished products. Today a lot of standard software products are released in short cycles having a new version every one or two years, sometimes even more frequently. This in fact has increased the productivity and the capability of software systems through the reuse of old software products (prototypes) for new product lines. But this has also led to increased maintenance costs especially through the frequently needed updating and upgrading of software and (due to the additional resource requirement of the new software) even hardware on the client side.

3 Service-oriented Computing

Another major difference between software products and industrial products is that the development of software products is the most cost intensive part, while copying and shipping (especially if the latter is done via internet) is very cheap, with costs close to zero. This leads to the idea of reusing often needed parts of a software product in other software products, to reduce the development cost by avoiding the "multiple reinvention of the wheel". Software developers identified this situation decades ago. The key approach is to ease the reuse of software components by a framework or component model. The idea of *components* is to package functionality regarding a specific solution into one easily usable program block, which is, for example, bundled in a so-called programming library. To ensure reusability it is important to define a well-structured *interface,* which is used by the application program or software to gain access to the component. The usage of components reduces not only the programming efforts by reducing the programming work, it also reduces the specification and planning costs as well as the general project risk. The reason for this is that components provide a higher level of abstraction to certain low level problems. For example a file handling component provides functionality to handle files easily on an abstract layer without forcing the developer to deal with low level file and storage related problems like file representation, directory representation, block allocation etc. Further, if components have already been used in other projects their reliability is already proved and possible pitfalls have been identified and maybe solved. This reduces the overall project risk.

Another positive aspect of the component model is that they in fact opened the market for low-level standard components like operating systems, graphical toolsets and other library-based products. Due to the ubiquity of communication networks, new technologies have been developed which allow the shared use of components by several connected machines. Well-known representatives of those technologies are Remote Procedure Call (RPC) (TANENBAUM, 2001), Common Object Request Broker Architecture (CORBA) (OBJECT MANAGEMENT GROUP, 2004) and the relatively recent Simple Object Access Protocol (SOAP) (WORLD WIDE WEB CONSORTIUM, 2003) that standardizes message exchange between related partners in the context of web services. Those distributed technologies enabled new components, which are used by several and often different applications within one enterprise. Today enterprise wide applications heavily depend on distributed technology, which in many cases leads to the following layered abstract structure of components:

1. *Operating Systems* provide basic functionalities, which needs to be locally present on all machines like file systems, memory and network management.
2. *Database Management Systems* store data in an application independent manner and provide high-level data manipulation and efficient data access functionality.
3. *Workflow Management Systems* are responsible for the process-based coordination of the execution of individual applications or functionalities.
4. *Applications* are software components which provide functionalities which are directly coupled with the enterprise business model.
5. *Graphical User Interfaces* support a proper representation of the applications on a user's machine and are responsible for the interaction of the application with the user.

While these technologies from a theoretical point of view allow the usage of components in an intra-organizational as well as an inter-organizational manner, most business models stuck for traditional, policy and security reasons on an intra-organizational usage. But the efficiency advantages of the use of distributed technologies for a fully automatic integration of inter-organizational business processes are obvious. Especially in the context of supply chain management, new concepts have been developed to automate the transfer of data between enterprises. The key idea of Electronic Data Interchange (EDI) is to replace paper documents, which are exchanged between companies, by digital documents. These digital documents contain in principle the same data as the paper documents.

To ease the exchange of the data, standards have been introduced like UN/EDIFACT or ANSI ASC X12. These standards prescribe the formats, character sets and data elements used in the exchange of documents, such as purchase orders and invoices. In order to use EDI for an automatic data interchange, two enterprises have to agree on a standard and on the meaning of the data. This usually means that before an EDI connection between two partners is operational, the partners have to go through a long and in many cases costly negotiation, contraction, and software adaptation process. This renders EDI connections only profitable for long running business connections where both partners have proved their creditability beforehand. As a result, this high start-up effort and standardized formats make EDI too expensive and inflexible regarding ad-hoc changes of partners or business models.

The recently introduced concept of service-oriented computing addresses the problem of making automatic – or electronic – data interchange more flexible and dynamic. One conclusion, which can be made from EDI, is that the transfer of data always invokes some kind of processing which leads to the execution of a service on the addressees' side. The service may be abstract, e.g. a transfer of money from one bank account to another, but the service may also me tangible, like the production of a specified product including the shipping of the product to the recipient, often the sender of the EDI message.

As a result, not only the transferred data is relevant, but also the service itself needs attention. The key idea of *service-oriented computing* is to provide well-described services to the market, which can be invoked by a published interface. In this context a *service* is a well-specified unit of work offered by a provider, which can be performed on demand. To enable the customer to choose the right service, a service needs to be specified sufficiently precisely. The resulting *service specification* is a complete characterization of what the service does, but the specification does not include how this is actually done. There are two reasons for two levels of abstraction: Firstly, the customer of a service is interested in what is being achieved by the service, not in the implementation details on how exactly something is actually achieved. Secondly, information about how a service is done might be a valuable company asset and as such, subject to intellectual properties rights management. In the provider side, the service implementation has to be made available to the customer through an interface. The *service implementation* is an executable representation of a procedure to fulfil the service specification. This representation says how a service has do be done and how it can be performed and by which resources, where resources include computer hardware and human involvement.

To make the service-oriented computing applicable, the *Service Oriented Architecture* (SOA) (BURBECK, 2000) has been proposed. (Figure 2) SOA defines roles and interactions of three types of participants which are crucial players in service-oriented applications:

1. *Service Providers* offer services to service requestors. They hold the service implementations including the needed resources for the execution of the services offered. They publish specifications of the offered services by means of the service brokers.
2. *Service Brokers* have the role of yellow pages, i.e. they act as a repository for many service specifications and the corresponding service providers. In addition, they offer powerful searching and matching tools to facilitate efficient finding of services by service requestors.
3. *Service Requestors* are the customers and they use the service brokers to find a suitable service which they invoke from the provider through an interface.

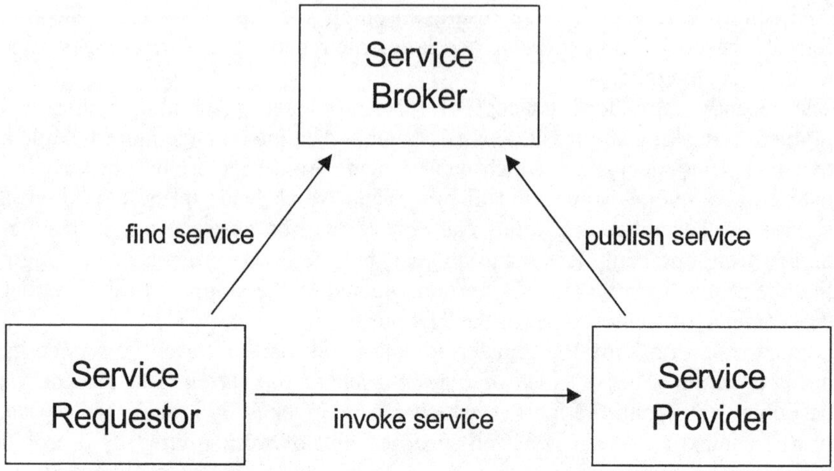

Fig. 2. Service oriented architecture

This architecture enables an automated ad-hoc invocation of services if computers can process the service specifications and the interfaces for service invocation. *Web services* are a current implementation of SOA. They provide computer readable service specifications and a standardized interface for service invocation based on web technologies like the Hypertext Transfer Protocol (HTTP) (WORLD WIDE WEB CONSORTIUM, 1999), the Extensible Mark up Language (XML) (WORLD WIDE WEB CONSORTIUM, 2004) and the Simple Object Access Protocol (SOAP) (WORLD WIDE WEB CONSORTIUM, 2003).

The vision for service-oriented computing in general and web services in particular can be described like this: Customers and companies get their demand quickly, easily and in an ad-hoc manner satisfied by electronically querying service brokers for required services. The service broker delivers a list of matching services and the corresponding service providers. This list may include payment information, but also electronic on-line negotiation with service providers should be possible. Depending on their business model, the service brokers will electronically charge an amount for each query; alternatively they may finance themselves by sponsored advertisements, for example similar to the business model of web search engines today. Based on this list of service specifications, the customer will choose the most suitable service provider and will electronically invoke the service for which the provider electronically charges them. Notice that the term suitable is highly case-specific; in general it might be the cheapest, most secure or fastest service or whatever criteria are of importance for the customer at that moment. This approach has the potential to create a new digital market. This market will have the following features:

- There are a large number of producers and consumers in the market. In theory the entire networked world can participate in this market.

- Results of service (as far as information is concerned) can travel freely through the computer network at high speed.
- The transaction costs will be close to zero. Especially if the service brokers will be paid by advertisements.

While this vision is striking, the software industry today faces severe technical and organizational problems that have to be solved in order to realize this vision. Some issues are of technical nature such as security issues, electronic (micro) payment and non-repudiation, which can be addressed by current technologies like encryption and signing in cooperation with adequate legal regulations. Other problems have an organizational or business-related flavour, especially regarding the needed investments in software and hardware, as well as the question who will make the first move and who actually wants to sell their products in nearly perfect competition provided by powerful service brokers. Finally there are still some conceptual problems that cannot be solved by pure technology or pure business solutions and concepts. Interestingly, potential solutions to these problems have their roots in philosophy and linguistics; we will sketch recent efforts in this direction in Section 5 of this paper.

It is also noteworthy that in the future service-oriented computing might have a strong influence on how software systems will be developed and on how the business models of large parts of the software industry will look. We will explain this shift by the following metaphorical analogy. If a software product would be a haircutter, then software industries most used business model is to sell (license, respectively) haircutters to the customers or enterprises. If a customer or enterprise wants to "use" the purchased haircutter to get his hair or the hair of his employees cut he has also to buy a fully equipped hair studio (which is equivalent to computer hardware in the software scenario). This situation is not cost-efficient if the customer does not need a hair cut very often. This is exactly the problem everybody has today when they need special software for the execution of a rarely used task. In a service-oriented world the solution for this uncomfortable situation is clear, the customer takes a look at the yellow pages to find a service provider providing the "hair cut service". The provider has already a haircutter and a fully equipped hair studio, which he makes available to the customer. The customer just needs to invoke the service on demand and he is charged per use. Carried forward to the software scenario, this results in a world where service providers supply computer hardware with software providing some services. These services can be invoked on demand and their usage will be charged per use. In such a world it is possible to think of software components, which are represented by services. These services may provide simple functions like file storage but also complex functionality like databases or applications. So software development could shift in parts from basic programming to the composition of several services to a resulting complex software product.

4 Process-oriented Information Systems

Since the early 1990's there has been considerable effort in developing a new type of middleware, dedicated to the management of business processes. These efforts have their roots in business process orientation, coined by the seminal work of HAMMER and CHAMPY. Their work is based on the observation that companies conduct their work in processes, where each process consists of a set of interrelated activities, which are performed by the co-workers of the business organization. Hammer and Champy have shown that by putting the focus on processes and by adapting the organizational structure of the company accordingly, massive improvements of the company's efficiency and effectiveness can be realized.

The middleware to facilitate the modelling and controlled execution of the automated parts of business processes is known as workflow management systems. By extracting the business process structures from applications, these processes suddenly become the subject of discussion and improvement. By providing graphical user interfaces for process modelling and by creating an immediate relationship between the graphical process structures and the workflow system, the improved processes can be realized with little effort, controlled by workflow management. In workflow technology, we use workflow languages to model workflows including their technical and organizational environment. A workflow language provides notation for expressing different dimensions of the problem domain; these dimensions are also known as workflow aspects. The main workflow aspects are the following:

- *Functional aspect:* The functional aspect characterizes what has to be done during the workflow. This aspect subsumes' functional decomposition, i.e., the breakdown of work units from complex processes (e.g., Customer Relationship Management processes) to individual activities (e.g. sending an offer for a new product). To this end, activities represent units of work that have to be enacted according to a specific process structure.
- *Behavioural aspect:* The behavioural aspect describes when and under which conditions activities have to be executed in the context of a workflow process. Important components of these aspects are control flow constraints and start conditions for the activities. The control flow constraints represent the execution order of activities while start conditions are used to define preconditions of a particular activity or an entire part of the workflow schema.
- *Information aspect:* The information aspect defines which data will be used, created or modified. An important aspect of workflow management is modelling the structure of application data. This is needed to allow the system to control the transfer of application data as generated or processed by workflow activities. While the basic principle of the information aspect is straightforward, there are many technical issues to solve, for instance different data format of the application programs, which have to work together. In this case, filters or converters are required to allow a seamless integration of different applications using different data formats.

- *Organizational aspect:* The organizational aspect determines who will perform an activity. A major goal of workflow management is enhancing efficiency by assigning work to persons or software systems specified in workflow schemas. To reach this goal, a workflow system has to be provided with information on the organizational and on the technical environment. Since a strict assignment of activities to persons is not feasible in most cases, the role concept is used. Before the start of an activity the workflow management system selects one or more persons, which are permitted to execute the activity. The activity will be put on the work list of those persons. After one person has initiated the execution of an activity the activity will be purged from the work list of the other persons.
- *Operational aspect:* The operational aspect defines how a workflow activity is being executed, i.e. which software system or software component will assist at the execution of a particular activity. The integration of existing tools, components and application programs into workflow applications is an important feature of workflow management systems. This aspect covers mainly technical issues, like the run-time environment, the definitions of input and output and their mapping to the respective workflow activities.
- *Flexibility aspect:* The flexibility aspect specifies what kind of changes can be applied under which conditions. One form of flexibility is the ability to change a workflow schema without re-coding the application programs, simply by restructuring the activities and their execution constraints. Another form of flexibility is the possibility to react to unforeseen situations. This can range from the re-executing of already executed activities to the execution of activities which were not planned in the workflow schema.

The design of business process models and their technical counterpart, workflow models is a tedious and often costly task. Similar to software development, workflow application development can be guided by methodologies. However, since the focus of a workflow project is the creation of a process-oriented workflow application, the focus is rather different; it can be sketched by the following phases:

- *Information Gathering*: During the information gathering phase, the information on the business processes is assembled by interview, observation, documents and business policy analysis.
- *Business Process Modelling*: The informally specified information on the business processes is aggregated using business process modelling languages; these languages provide graphical notation for activities, ordering of activities and other constraints between activities. Depending on the language used, additional concepts can be specified. The result of this phase is a business process model.
- *Business Process Model Improvement*: Domain experts improve the business process as specified in the model. Criteria for improvements include higher customer satisfaction, for instance, by providing faster and more flexible services to the customer and reduced cost of operation. Efforts in this phase are

also known as business process re-engineering or business process optimization. This phase might also involve the re-structuring of the organization of the company, i.e. modification not only of the processes but also of the organization that enacts them.

- *Workflow Modelling and Implementation*: Once a business process model is defined, it needs to be implemented. The selection of specific middleware supporting processes, for instance workflow management system, has to be done. Using the notation and language provided by that system, the parts of the business process model that can be automated are implemented. Just like in traditional software engineering, various tests and prototype development is suitable.
- *Operational Phase*: In the operational phase, the business process instances, the concrete customer processes are enacted. The workflow system logs execution information carefully so that it can be used in future iterations of business process model improvement.

The advantage of process-aware information systems, such as workflow management systems, is the efficient and effective enactment of workflows according to a given specification. As a result, delays of work can be minimized and bottlenecks of a process can be detected. For instance, if a particular co-worker with specific competences is overloaded then by hiring a second person with this qualification might lead to considerable improvements in the response time of the overall process. In addition, by communicating the state of execution of a process to the customer, she can be informed timely on the current state of her process. For instance, an insurance claim process may currently wait for an external expert, or the financial settlement might have just been arranged. This valuable information can be communicated to the customer, which enhances customer satisfaction and improves transparency of doing business.

While there are numerous success stories in workflow management, there are also application domains in which the fixed structure of the process is not adequate. In these settings, advanced workflow concepts have to be installed, such as the dynamic change of business processes to process specifications and the possibility to allow running instances to follow a new and improved process model, also known as process instance migration (VOSSEN and WESKE, 1999; WESKE, 2000; REICHERT and DADAM, 1998).

A recent trend in workflow computing is the integration of the processes of several companies into one inter-organizational process to enhance the efficiency of business-to-business scenarios (VAN DER AALST and WESKE, 2001). These efforts fit very well into the current endeavours in service-oriented computing sketched above because for an efficient implementation of such interorganizational processes, an electronic data interchange is indispensable. The core idea is to make the interfaces between companies visible in the workflows to analyze them regarding timing, usefulness and costs. Consequently the next step is to reduce media breaks by the usage of electronic data interchange and to reduce manual handling of interface data where it is not essential for the business that the data is pre-processed manually. Finally the integration should and will not stop on the

the data exchange level, but it will be continued up to the workflow control and monitoring level. This means that it is possible to monitor the state of running inter organizational workflow instances even when the current task is executed by resources of another company. Logically the following step is to use this infrastructure to monitor deadlines in an inter-organizational way to reduce spreading of throughput time and to have more control on the bullwhip effects (LEE, PADMANABHAN and WHANG, 1997) of the whole inter-organizational workflow.

5 The Future: Semantics, Ontology and Planning

Current efforts in establishing service-oriented computing have not only to deal with technological and business-related problems, which are a usual factor in the information systems discipline, but also with philosophical and linguistics related challenges (refer to section 3). These challenges have their roots in the question of what is the semantics of the transmitted data. By semantics we mean here the meaning of words and expressions and their relationship to other words and expressions not only in the computational context, but also in the real world, the human and natural language related context. This raises the following question: Why is semantics a problem that is not grave enough to be perceived when dealing with inter-organizational software and what changes when moving from inter to intra-organizational software and especially to service-oriented computing?

We will start with the first part of the question. When programmers are programming classical, non-distributed software from the scratch without reusing any closed-source libraries, like in the early phases of computing, the problem of semantics is not evident. Questions like "What is the meaning of the input parameter named x of type y of procedure z?" can be answered even if no documentation is available simply by looking into the source code of the concerning procedure. The problem of semantics worsens a little bit when we take the next step, but it is still not really striking. The next step is software development by means of closed source libraries and distributed components. Here, in fact, good documentation is unavoidable to get the meaning of certain parameters and calls, but humans process the documentation during programming. Humans are able to understand the documentations in most cases by means of their background knowledge on the world in general and information systems or the concerning business models in particular. Even if they are not able to find the correct semantics, because the documentation is too bad, they usually have the possibility to ask the vendors or experts on the components by mailing-lists, e-mail, phone, etc. This situation does not even change when we look at traditional (not service-oriented) electronic data interchange. There are still humans who are programming the interfaces, which make two inter-organizational systems working together, but it is worth mentioning that the semantics problem gets really tangible, even if it is still solvable in the old fashioned manner. Experience shows that the negotiation process, which is needed to establish the electronic data interchange, can be lengthy. This is because agents and developers of the different companies often have different backgrounds and experiences which cause understanding problems when going into details. So the involved partners have to find a common language with common semantics to describe the relevant concepts which are behind the exchanged data structures in detail.

All this is not possible in service-oriented computing. Since the goal of service-oriented computing is to automatically find and invoke proper services without the need of human assistance. This means that the involved computer systems have to find and agree on a common language including semantics to describe what a service is about and what the customer expects from a service he is looking for. This is currently and presumably in the near future not practicable with computing technology. So the answer for the second part of the above question is the change from traditional computing to service-oriented computing is the elimination of the human being as the mediator for the semantics of the exchanged data between software components. To implement service-oriented computing we have to avoid a situation in which computers have to build up and negotiate semantics from scratch, rather we have to provide a formalized semantics which can be used by different computer systems to pin down the semantics of exchanged data and the preconditions and effects of a service. This is the point where ontologies come into play.

Ontologies

The origins of the term ontology are located in philosophy. Later the term has been modified and used by several scientific branches like computer science, but the foundation of the term is still located in philosophy. For this reason we want to start with the philosophical definition of this term: Ontology as a branch of philosophy is the science of what is, of the kinds and structures of objects, properties, events, processes and relations in every area of reality. (SMITH, 2002) The goal of the philosophical ontology (as a science) is to fathom the truth. (ZÚÑIGA, 2001) For this reason in the context of philosophy the term ontology has to be used in singular form, because only one "true" truth may exist.

During the adoption process from philosophy to computer science the term ontology has sensed a smooth change in its definition. Two things can be stated: First, there is (currently) no consensus regarding the definition of the term ontology in detail. Second, a consensus exists, that an ontology in computer sciences is a representation of domain specific knowledge written in a formal language. (ZÚÑIGA, 2001) So it is clear that there are two differences existing between the philosophical ontology and the computer science ontology: An ontology in computer science is not a scientific branch and it is pragmatically related to an application domain. Further it is possible to use the term in the plural form, because several representations of domain specific knowledge may exist. (KUROPKA, 2004)

A known and often used definition of the term ontology in computer science context is the definition deployed by GRUBER: "An ontology is an explicit specification of a conceptualization." (GRUBER, 1993; GRUBER 1995) One major weakness of this definition is the fact that regarding this definition nearly every idea, which has been written down in a formal manner (e.g. a computer program) is an ontology. For this reason we will use here the following tighter definition: An ontology is a model of linguistic means of expression on which several actors have agreed on and which are (or can be) used by those actors. (KUROPKA, 2004) It is worth mentioning that in this context a model is a representation of an (not necessary tangible) object system written by the mean of a formal (but not necessary textual) modelling language.

So ontologies can be used to pin down the common understanding, the semantics of words and concepts. If two actors are sharing the same ontology they can directly start to communicate about relevant issues without going through a lengthy negotiation process on common understanding. Ontologies written in formal languages, which are process-able by computers, are a possible solution to the challenges of fully automated service-oriented computing in general and web-services in particular. If partners of a service-oriented market agree on one particular ontology as well as one formal language for the description of service specifications, then we have a semantic service specification. This semantic service specification facilitates an automated mediation of the semantics between software components, which makes the vision of service-oriented computing feasible. For example, if a service requestor requests a service, which is to ship a power socket, the result of the service invocation may be different when two different service providers are involved and no commitment on a common ontology exists. The reason for this mishap is that a power socket of a German provider is different to the power socket of a British provider. A common ontology solves this problem, because it provides a common view on the world. In this example the ontology should have information about the fact that in different countries different power sockets are in use. In Germany the SCHUKO power socket and in Britain the BS power socket. This causes naturally a different specification of the "ship power socket service" for both providers. The German provider for example will only ship SCHUKO power sockets while the British provider will only ship BS power sockets. So requesting a list of providers shipping power sockets in general will have a different result. The result depends on the "intelligence" of the query mechanism. It can contain an empty list or it can contain both providers including a hint about the fact that different socket standards are existing and that the requestor has to choose one standard. It can even contain a proposal on the right provider for the requestor, if the system is able to use the information about requestor's nationality properly.

The commitment of several partners to an ontology is some kind of a standardization process. As always, making standards is not a simple thing, but as already shown in other contexts it is practicable if one strong partner is involved (for example refer to EDI standardization in the automotive industry) or if something new is invented and its adoption is very rapid (e.g. TCP/IP (DEFENSE ADVANCED RESEARCH PROJECTS AGENCY, 1981) and the other Internet protocols like HTTP (WORLD WIDE WEB CONSORTIUM, 1999), SMTP (POSTEL, 1982), etc.). Currently great endeavours are invested in the development of powerful and computer process-able languages for the description of ontologies. Some examples are KIF (GENESERETH and FIKES, 1994), GOL (DEGEN, HELLER,

HERRE and SMITH, 2001), DAML+OIL (DEFENSE ADVANCED RESEARCH PROJECTS AGENCY, 2001), RDF (WORLD WIDE WEB CONSORTIUM, 2004b) and WSML (DE BRUIJN, 2004). On the basis of those languages a bunch of ontologies is currently under development, for example, general purpose ontologies like CYC (http://www.cyc.com) and OpenCYC (http://www.opencyc.org) or service-oriented ontologies like WSMO (ROMAN, LAUSEN and KELLER, 2004).

Automated Process Planning

Adding semantics to the concept of service-oriented computing by means of using an ontology is an important step forward to create a new digital market with full information and negligible transaction costs as envisioned in Section 3. In the context of service oriented computing a semantic service specification enables the automation of a variety of reasoning tasks such as discovery, selection, invocation, composition and monitoring of services (NARAYANAN and MCILRAITH, 2002). Semantic service specification means to give a description or specification of the functional and non-functional properties of a service in a logical, machine-interpretable language on the basis of a formal ontology that the involved partners have committed to. This comprises of a description of the preconditions under which a given service can be executed and the effects that the execution of this service will have. In addition to this functional description, non-functional properties can be specified, e.g. duration or costs. If service providers specify and publish services using semantic service specifications, the discovery of services can be automated as exposed above. Therefore, service requestors describe their functional and non-functional requirements on the requested service in a logical language, e.g. a requestor asks for the cheapest service that returns a creditability report for a customer within 2 minutes or that invokes production and shipping of a certain product within a week. Due to the machine-interpretable specifications of the provided services and the requested service, discovery can be automated. As a result, a vast amount of offers can be searched and compared in little time with negligible costs. Thus, semantic specification of services constitutes a big step forward to the digital market.

A crucial point for the competitiveness of organizations is the performance of their processes (HAMMER and CHAMPY, 1993). Workflow Management systems are supporting companies to handle this challenge by automating and supporting the enactment of processes. However, planning, modelling and supplying the workflow schema to the workflow management system is usually done manually. Manual planning involves the time intensive work of specialists and thus is a costly task. For this reason, it is not economically advantageous to plan an individual workflow schema for every business case. Instead, similar cases are grouped together and planned conjointly. For example, there is only one generic workflow schema used for all incoming purchase orders. Finding a workflow schema that copes with all possible variants of purchase orders becomes more difficult the more the structure of the cases varies. Normally, one workflow schema is used to control the enactment of a process, but if something unanticipated occurs threatening the achievement of the goal, it becomes necessary to react by changing the process. This re-planning has to be done manually and thus delays execution. Afterwards the workflow management system has to adapt the process enactment to the changed workflow schema as discussed in Section 4. This change of workflow schema at run time has already been investigated (ELLIS, KEDDARA and ROZENBERG, 1995; HAGEN, ALONSO, 2000; REICHERT and DADAM, 1998). It proves to be a difficult task for the workflow management system to assure consistency, because it does not have all the required information on the dependencies between process steps.

Current research approaches (BLYTHE and DEELMANN and GIL, 2003; LAZOVIK, AIELLO and PAPAZOGLOU, 2003; SCHUCHEL, WESKE 2004; ZENG, BENATALLAH, LEI, NGU, FLAXER and CHANG, 2003) evaluate options to use semantic description of services to enable automated service composition. Reasoning algorithms like Artificial Intelligence planning algorithms (GHALLAB, NAU and TRAVERSO, 2004) are used to generate workflow schemas. Inputs for a planning algorithm are the semantic descriptions of the provided services and the requested service. A planning algorithm does not only check if a single service matches the request, but it also tries to find a set of services combined in a workflow schema to satisfy the request. Workflow management systems can then take the generated workflow schema as input to support and automate process enactment and schedule the single services. Since automated service composition is done without human effort, it changes the possibilities to organize workflow planning and enactment. Instead of planning just one workflow schema for a group of similar cases, automated planning allows the individual planning of every case. Thus, an optimized workflow schema that takes all the specific properties of the case into account is generated. This capability can realize a competitive advantage, since individual treatment of cases and thereby of customers becomes increasingly important (HAMMER and CHAMPY, 1993). Furthermore, replanning can be triggered and performed automatically if necessary. Thus, a delay of the process execution can be avoided. If a workflow schema has to be changed at runtime, consistency can be assured, because all necessary information on the dependencies between the process steps is available for the planning system.

The ability to automatically discover and compose services to match the needs of service requestors, impressively demonstrates the potential of semantic specification of services. In addition to service discovery and composition it allows automating service invocation, negotiation on service properties, process simulation and monitoring. As a result, transaction costs are minimized and the consolidation of supply and demand is optimized.

6 Conclusions

This paper sketches fundamental concepts and recent trends in software, information systems and process technology. By joining the forces of information systems competences and business and economical competences, an added value can in fact be reached. In a networked world only by combining these complementary competences can the challenging questions that we face today can be understood better.

References

VAN DER AALST, W.M.P.; WESKE, M. (2001): The P2P Approach to Interorganizational Workflows. In: Dittrich, K. R., Geppert, A., Norrie, M. C. (Eds.) Proceedings of the 13th Conference on Advanced Information Systems Engineering (CAiSE'01), pp. 140-156. Springer Lecture Notes in Computer Science 2068. Heidelberg: Springer.

ANSI/IEEE Std. 723-1983 (1983): IEEE Standard Glossary of Software Engineering Terminology, New York, IEEE, 1983.

ANSI Std. (2004): ASC X12. http://www.x12.org/

BLYTHE, J.; DEELMAN, E.; GIL, Y. (2003). Planning for Workflow Construction and Maintenance on the Grid. In ICAPS'03 Workshop on Planning for Web Services, 2003.

BOEHM, B. (1988), A Spiral Model of Software Development and Enhancement. IEEE Computer (21), 1988, pp. 61-41.

DE BRUIJN, J. (2004), The WSML Family of Representation Languages. 2004. http://www.wsmo.org/2004/d16/d16.1/v0.2/

BURBECK, S. (2000), The Tao of e-business services. IBM Corporation, 2000. http://www.ibm.com/software/developer/library/ws-tao/index.html

DEFENSE ADVANCED RESEARCH PROJECTS AGENCY (2001), DAML+OIL (March 2001). 2001. http://www.daml.org/2001/03/daml+oil-index.html

DEFENSE ADVANCED RESEARCH PROJECTS AGENCY (1981), RFC 793 - Transmission Control Protocol. 1981. http://www.faqs.org/rfcs/rfc793.html

DEGEN, W.; HELLER, B.; HERRE, H.; SMITH, B. (2001), GOL: A General Ontological Language. Proceedings of Formal Ontology in Information Systems, 2001.

ELLIS, C.; KEDDAR, K.; ROZENBERG, G. (1995), Dynamic change within workflow systems. In Proceedings of conference on Organizational computing systems, pp. 10-21. ACM Press, 1995.

GENESERETH, M.; FIKES, R. (1994), Knowledge Interchange Format Version 3.0. Computer Science Department, Stanford University, 1994.

GHALLAB, M.; NAU, D.; TRAVERSO, P. (2004), Automated Planning: Theory and Practice. Morgan Kaufmann, 2004.

GRUBER, T. (1993), A Translation Approach to Portable Ontology Specifications. In Knowledge Acquisition (5), 1993, pp. 199-220.

GRUBER, T. (1995) Toward Principles for the Design of Ontologies Used for Knowledge Sharing. In International Journal on Human-Computer Studies (43), 1995, pp. 907-928.

HAGEN, C.; ALONSO, G. (1993), Exception Handling in Workflow Management Systems. IEEE Trans. Softw. Eng., 26(10): pp. 943-958, 2000.

HAMMER, M.; CHAMPY, J. (1993), Reengineering the corporation. Harper Collins Publishing, New York, 1993.

KUROPKA, D. (2004), Modelle zur Repräsentation natürlichsprachlicher Dokumente: Ontologie-basiertes Information-Filtering und -Retrieval mit relationalen Datenbanken. Berlin, Logos, 2004.

LAZOVIK, A.; AIELLO, M.; PAPAZOGLOU, M. (2003), Planning and Monitoring the Execution ofWeb Service Requests. In 1st international conference on service-oriented computing (ICSOC'03), Trento, 2003.

LEE, H.; PADMANABHAN, V.; WHANG, S. (1997), Information distortion in a supply chain: the bullwhip effect. Management Science, Volume 43, Issue 4, 1997, pp. 546-558.

NARAYANAN, S.; MCILRAITH, S. (2002), "Simulation, Verification and Automated Composition of Web Services," in Proceedings of the 11th International World Wide Web Conference, 2002.

OBJECT MANAGEMENT GROUP (2004), Common Object Request Broker Architecture: Core Specification. Version 3.03. 2004. http://www.omg.org/docs/formal/04-03-01.pdf

POSTEL, J. (1982), RFC 821 - Simple Mail Transfer Protocol. 1982. http://www.faqs.org/rfcs/rfc821.html

REICHERT, M.; DADAM, P. (1998), ADEPT flex -Supporting Dynamic Changes of Workflows Without Losing Control. Journal of Intelligent Information Systems, 10(2): pp. 93-129, 1998.

ROMAN, D.; LAUSEN, H.; KELLER, U. (2004), Web Service Modeling Ontology (WSMO). 2004. http://www.wsmo.org/2004/d2/v1.0/20040920/

ROYCE, W. (1989), Managing the Development of Large Software Systems: Concepts and Techniques. Wescon Technical Papers, Western Electronic Show and Convention, Los Angeles, 1970. Reprinted in: Processings of the 11th International Conference on Software Engineering, Pittsburgh, 1989, pp. 328-338.

SCHACH, S. (1993), Software Engineering. Second Edition. Homewood, Irwin and Aksen, 1993.

SCHUSCHEL, H.; WESKE, M. (2004), Triggering Replanning in an Integrated Workflow Planning and Enactment System. In 8th East-European Conference on Advances in Databases and Information Systems, Vol. 3255 of LNCS. Springer, 2004.

SMITH, B. (2002), Ontology and Information Systems, 2002. http://ontology.buffalo.edu/ontology%28PIC%29.pdf

TANENBAUM, A. (2001), Modern Operating Systems. 2nd Edition, Prentice Hall, 2001.

UNITED NATIONS (2004), United Nations Directories for Electronic Data Interchange for Administration, Commerce and Transport. http://www.unece.org/trade/untdid/welcome.htm

VOSSEN, G.; WESKE, M. (1999), The WASA2 Object-Oriented Workflow Management System. In Proceedings of the 1999 ACM SIGMOD Conference on Management of Data, pp. 587-589. New York: ACM 1999.

WESKE, M.(2000), Workflow Management Systems: Formal Foundation, Conceptual Design, Implementation Aspects. Professorial dissertation, University of Münster, 2000. http://bpt.hpi.uni-potsdam.de/lib/papers/habil.pdf

WORLD WIDE WEB CONSORTIUM (2003), SOAP Version 1.2 Part 0: Primer. 2003. http://www.w3.org/TR/2003/REC-soap12-part0-20030624/

WORLD WIDE WEB CONSORTIUM (2004) Extensible Markup Language (XML) 1.1. 2004. http://w3c.org/TR/2004/REC-xml11-20040204/

WORLD WIDE WEB CONSORTIUM (2004b) RDF/XML Syntax Specification. 2004. http://www.w3.org/TR/rdf-syntax-grammar/

WORLD WIDE WEB CONSORTIUM (1999) Hypertext Transfer Protocol -- HTTP/1.1. 1999. http://w3c.org/Protocols/rfc2616/rfc2616.html

ZENG, L.; BENATALLAH, B.; LEI, H.; NGU, A.; FLAXER, D.; CHANG, H. (2003), Flexible Composition of Enterprise Web Services. Electronic Markets - Web Services, 13(2), 2003.

ZÚÑIGA, G. (2001), Ontology: Its Transformation From Philosophy to Information Systems. In Proceedings of Formal Ontology in Information Systems, 2001, pp. 187-197.

Digital Social Market Economy – Towards a New Economic System

Paul J.J. Welfens[1]

1 Introduction 25
2 Digital Social Market Economy: Expansion and New Problems 28
2.1 Digital Structural Change and Growth Bonus 28
2.2 Selected Specifics of a Digital Economy 29
2.3 Network Effects 30
2.4 Innovation Aspects 36
2.5 Digital Media Society 39
2.6 The Software Economy 40
2.7 Competition, Efficiency and Welfare-maximizing Differentiated Two-Part Tariffs 42
3 New System Design and Strategic Policy Aspects 50
Appendix: Product Innovations and Network Effects in a Simple Model 56
References 59

1 Introduction

The expansion of the information and communication technology (ICT) in the 1990s has raised new challenges for OECD countries and Newly Industrializing Countries. With a share of about 10% in the US, Germany and Scandinavia, the ICT sector is still rather small, but its sustained technological and economic dynamics clearly influence overall structural change in a decisive way. High US economic growth in the 1990s and some EU countries has largely been attributed to ICT whose expansion requires a high share of skilled labour and strong emphasis on innovations so that research and development (R&D) become more important. Since ICT facilitates production of many new varieties, represents network effects in certain sectors and stimulates innovation, it has those ingredients considered particularly important in the New Growth Theory.

In the Lisbon agenda of 2000, the EU declared that it wants to become the most competitive and dynamic knowledge-driven economy by 2010. Higher growth and a higher participation rate – hence more jobs – are expected from this initiative in which "open coordination", based largely on the incentives associated with benchmarking and political peer pressure, plays a prominent role. National governments and the EU have adopted various policy initiatives to promote expansion of the digital knowledge economy; e-government and e-commerce are expected to

[1] I gratefully acknowledge technical support and research assistance by Dora Borbély and Jens Perret, EIIW. The usual disclaimer applies.

grow strongly in the medium term. Falling prices of telecommunication services and in particular of narrow-band and broad-band internet use are expected to stimulate digital networking, innovation, diffusion and growth.

The economic system will have to adjust as the New Economy – the digital economy – is unfolding. Institutions and rules which were shaped in the age of industry are no longer adequate for the digital age. New rules and institutions can help to better exploit the productivity potential of ICT and to cope with some key problems of the digital economy.

The traditional social market economy – as developed in Western Europe in the 20th century – has rested upon three pillars:

- allocation of resources through competitive markets; this was the standard case except for the military sector and the case of the so-called natural monopoly for many years – the latter representing the telecommunications sector, the energy sector and the railways system. Competition will result in efficient production of goods and services where producers must take consumer preferences into account; accordingly, market economies are characterized by consumer sovereignty. The income generated in markets is, of course, not equal. Nevertheless, the market system in a broad sense provides equal opportunities. Consumers have a choice among competing firms. Those who have entrepreneurial ambitions can enter markets as a newcomer on the supply-side;
- redistribution through taxation, public heath care insurance and the public pension system and universal/public services which offer uniform prices across space (the latter requiring implicit cross-subsidization for both poor and rural regions).

This system has come under threat through economic globalization, associated with more foreign direct investment and economic opening up in China and eastern Europe, which has forced EU countries to reduce corporate tax rates in the 1990s. Expansion of digital trade also contributes to undermining the tax base of OECD countries so that national governments are forced to reduce public expenditures relative to national income. The only way to counter-balance this effect is through regional political integration, namely that governments of different countries – willing to harmonize taxes in some way or the other – effectively create a larger country so that a downward tax spiral is avoided.

Moreover, the increasing use of modern computers and software has created larger markets which basically call for more cross-border cooperation in competition policy. Otherwise the larger relevant markets might become less competitive. At the same time the adjustment speed in financial markets has increased strongly so that overshooting problems (short term deviation from long equilibrium value of the real exchange rate) have potentially become more important. Strong exchange rate overshooting which basically reflects differences in the fast adjustment speed of financial markets and slow adjustment speed in goods and factor markets can seriously destabilize a country's economy. While it is true that ICT raises productivity and output growth, it also could be argued that there will be more long term volatility of output and prices so that the average welfare effect of

ICT could be ambiguous in certain cases, namely when the increase in volatility more than offsets the increase in output growth.

ICT is considered to be quite important for productivity growth (VAN ARK, 2001; WELFENS, 2003). There is no doubt that technological progress in the production of ICT goods is high, and there are also good arguments for why the use of ICT goods contributes to higher productivity. ICT can help cope with the ageing of society and shrinking populations in OECD countries where growth normally should be expected to fall considerably in the 21st century. Increasing use of ICT and continuing progress in ICT production could stabilize growth rates. Physically difficult work – a potential problem in ageing societies - can be largely automated. The main work of the 21st century is clearly in the services sector, which came to dominate employment in most OECD countries already in the 1960s and 1970s. Modern software tools help even to computerize many services. This is the case, for example, in banking, education, and health care. Ageing could imply a less innovative society and hence lower growth; however, ICT dynamics and an adequate policy of recurrent training/retraining and digital upgrading of human capital could allow for increased productivity time and again in all age brackets.

Liberalization of telecommunications in the EU in 1998 as well as in Asia and Latin America in the late 1990s has stimulated competition, resulting in falling prices and in a wave of innovations. It also is noteworthy that Eastern European countries have quickly embraced a phased-in approach to liberalization of telecommunications which has become a dynamic field of expansion in countries once known for a large long term excess demand. As regards international competitiveness in sectors relevant for ICT production, EU accession countries have a competitive disadvantage (BORBELY, 2004), while the US, Japan, Korea, Germany, France, the Netherlands, Finland, Sweden and Ireland are relatively well positioned among OECD countries (OECD, 2003). In many poor countries – including NICs and LDCs – expansion of mobile telephony has been much faster than in fixed line telephony.

The digital economy is embedded in the information society in which information is a distinct input in the production function. Some of the well-known imperfections of information markets could thus become a crucial challenge for the design of institutions and rules of a market economy. Since a digital society is strongly based on networks, there is wide-ranging physical and economic interdependency which must be analyzed. In the following analysis, we look at selected characteristics and problems of the digital economy and then propose necessary institutional innovations and policy reforms which could help to establish a stable and prosperous economic system.

2 Digital Social Market Economy: Expansion and New Problems

2.1 Digital Structural Change and Growth Bonus

The expansion of the digital economy is driven partly by the liberalization of telecommunications markets (e.g., network operation and voice telephony in the EU in 1998). Tthe UK had already liberalized and privatized the telecommunications sector in the 1980s under the Thatcher government, while the US broke up the private monopoly AT&T ("Ma Bell") into regional baby bells and a new AT&T, with the latter playing a limited role in long distance and international calls and where it is facing competition from other network operators. Both increasing competition and technological dynamics have made telecommunications the field of patenting with the highest growth rate in the 1990s in the EU and the US.

The strong growth of ICT in the 1990s contributed to an expansion of communication networks – broadly defined – and hence to acceleration effects in productivity growth. Typically, the usefulness or real value-added (U) of a network increases with the square of the number (L') - which we will also identify as skilled labour - of its users, so that

$$U = L'^2 \tag{1}$$

By contrast, a simple Cobb-Douglas production function relevant for the traditional sector reads

$$Z = K^{ß} L^{1-ß} \tag{2}$$

where K is factor input, L unskilled labor input and ß an exponent where $0<ß<1$. Hence we have

$$g_Z = ßg_K + (1-ß)g_L \tag{2'}$$

In the presence of networks it holds – according to equation (1) – for the growth rate g_U (with g as symbol denoting growth rates) that

$$g_U = 2g_{L'} \tag{3}$$

The growth rate of usefulness of a network is twice as high as the growth rate of the number of network users. Hence, the growth of U is higher than under Cobb-Douglas. If aggregate output Y consists of two sectors, Z and U, we have (with Q defined as the relative price p"/p – with p" representing the price of U and p the price of Z – and a defined as QU/Y) aggregate output defined in terms of the traditional good as

$$Y = Z + QU \tag{4}$$

$$g_Y = a[g_Q + g_U] + (1-a)g_Z = 2a\, g_{L'} + ag_Q + ß(1-a)g_K + (1-ß)(1-a)g_L \tag{5}$$

With a given growth rate of K and an equality of $g_L = g_{L'}$ - a special case to be relaxed subsequently - there will be a rise in output growth as long as the share a=QU/Y is increasing.

More generally (and giving up the assumption $g_L = g_{L'}$), the variable a can increase as long as labour is shifting from the traditional sector to the digital sector [with a given overall number of workers L"=L'+L so that dL'/dt= -dL/dt and hence $g_L=-(L'/L)g_{L'}$]. Giving up the special assumption $g_L=g_{L'}$, we can state the growth rate of aggregate output strictly as a function of the growth rate of capital and of the weighted growth rate of skilled labour.

$$g_Y= ß(1-a)g_K + [2a-(1-ß)(1-a)(L'/L)]g_{L'} + ag_Q \quad (6)$$

The net effect of both structural change and the relocation of workers is (ignoring g_Q) positive if

$$\ln2+ \ln a +ß +a > \ln(L'/L) \quad (7)$$

That is, the natural logarithm of the ratio of skilled labor to unskilled labor must be smaller than ln2 + lna + a +ß; this is, of course, trivial for L'<L. Hence as long as the share of unskilled workers is smaller than that of skilled workers, expansion of the skill intensive sectors always contributes to a higher growth rate.

If relocation of an unskilled worker L to the digital sector is associated with a required skill upgrading effort γ ($0<\gamma<1$) one should further take into account that $dL'/dt = (1-\gamma)dL/dt$ where γ represents the fraction of people working in the retraining sector. The share a(t) of "networked digital output" U can be assumed to depend positively on the rate of fall in the ratio of price p" of ICT goods to standard capital goods p, so that a=a(p"/p); with Q=p"/p and $\partial a/\partial[p''/p]<0$.

In a digital economy, there will be a faster diffusion of knowledge and – in combination with rolling out ever-faster digital networks – there will be a rising supply of innovative digital services. The latter could include increasing international digital outsourcing. Faster international (and national) diffusion of knowledge implies that high wage countries with dominant technology-intensive export will be less able to earn high Schumpeterian rents in global markets so that high wage countries will face increased pressure to move up the technological ladder and to focus more on product innovations for which higher prices can be fetched in world markets.

2.2 Selected Specifics of a Digital Economy

We have already seen in a simple model that there could be a temporary acceleration of aggregate real income growth. However, there are certain traits of the ICT sector which makes a long term growth bonus rather uncertain. A key aspect of a digital services economy is that the majority of workers are actively involved in generating, storing and processing information. However, information markets are rather imperfect, since the supply side must reveal part of information for free in order to let the buyer assess the value of the information good. Thus it is difficult to recover an adequate price for information in the market. Main elements of the digital economy are thus as follows:

- Information markets are imperfect (Arrow): There is a natural tendency for firms to swap information or rely on intra-company transactions. In electronic networks, information can easily be disseminated and stored, with marginal costs close to zero. However, creating new useful information and storing it to esnure that new knowledge becomes accessible requires the use of many scarce resources. The very fact that diffusion of new knowledge is rather easy through digital networks makes it difficult for creators of new knowledge to earn adequate income from "breeding knowledge". Hence digital intellectual property rights are quite important.
- Information is a confidence goods, and as such, building reputation and bundling become important. As regards bundling, the idea is to combine selling of a normal good (e.g., hardware: inspection good) and a confidence good.
- Bundling plays a crucial role in the selling of software (BAKOS and BRYNJOLFSSON, 1999; 2000). Software represents a system of subprograms, each containing algorithms to solve a subproblem, whose usefulness differs across potential customers. By creating software packages, software firms are able to fetch a rather average (high) price in the market while incurring only minimal costs in terms of added programming for a complex programme.
- There are problems of sustained competition in telecommunications and in software markets. Economies of density in fixed network telephony create a tendency towards a certain degree of dominance by a single operator in the access market. Economies of scope and economies of bundling can stimulate vertical integration which then creates the problem of transfer of market power from one market to another market.
- The role of the communications infrastructure for the digital economy is high. The faster, the more ubiquitous and the cheaper the use of networks, the larger the network effects and the greater the opportunities to launch digital product innovations. Modern regulation allows for the imposing of pricing regimes which create an incentive to achieve efficient output in network industries (here, telecommunications).

2.3 Network Effects

In principle, network effects are an endogenous outward rotation (or rightward shift) of the demand curve. Esentially, network effects occur in markets in which users of goods and services interact systematically. For example, if all use the same telecommunication standard, interconnection is easy and increases the usefulness of being linked to the telecommunications network. If all partners in a project group have access to fax (or SMS or MMS), there are better opportunities for coordination. Correct and adequate anticipation of network effects is not always easy, as we know from the fragmented patches of telecommunications networks in the UK in the 19th century when telecommunication operators finally called for consolidation through nationalization of the network. A single national operator could find it rather easy to establish regions with mutually-competitive networks, and it also would be able to anticipate network effects with a low degree of uncertainty. However, having a single firm creates monopoly problems and hence static and dynamic inefficiency. In the case of network effects, there is a large difference between monopoly output and competitive output. As such, the welfare loss from monopolization is large. Conversely, adequate regulation which forces firms to follow the rule marginal costs = marginal willingness to pay – or as a second-best solution: average costs=marginal willingness to pay – could yield enormous benefits for both the consumer and society.

The following graph shows an initial demand curve (DD_o) and the corresponding marginal revenue curve (R'_o) for the case of a natural monopoly (with falling marginal costs k' and falling average costs k). Network effects are shown as an outward rotation of the demand curve. It is drawn in such a way that the original demand curve coincides with the new marginal revenue curve. If there is a monopoly, there will be an increase in output from q_o to q_1, while price falls modestly. However, if there were competition – with firms assumed to set prices according to average costs - in the long run, there would be a large increase in output and a strong fall in prices. If the market considered were a market with production innovations, there naturally would be an initial monopoly position; in the diffusion stage – with newcomers entering the market –, however, the competitive solution could be realized.

Welfare analysis suggests that the optimum output is q_2, characterized by equality of marginal costs and marginal willingness to pay. However, this solution can be realized only if government pays a subsidy. Paying subsidies raises serious problems, particularly if the innovation rate in the industry considered is rather low. Many other sectors would also desire subsidies, and thus considerable risks of economic inefficiencies would occur.

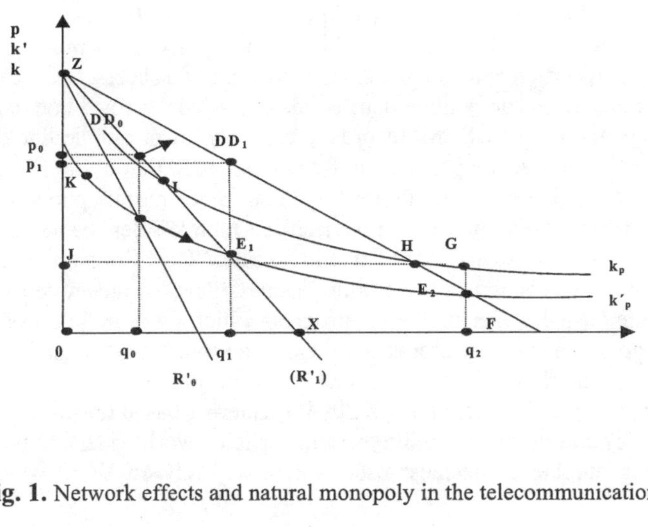

Fig. 1. Network effects and natural monopoly in the telecommunications sector

If the regulator is able to force firms to basically consider long run marginal costs as well as an add-on element in price setting so that effectively average costs are covered by market prices, society will come close to the efficient outcome q_2. Alternatively, firms can use two-part tariffs where one element basically is designed to recover fixed costs while the second element covers marginal costs. The regulator could then only focus on the second pricing element. At the same time the regulator could fix a basket of services for which an RPI-X-factor will apply. Based on the estimated sectoral productivity growth, X, the relative price of a basket of services must be reduced continuously. The growth rate of the price basket is expected to fall according to inflation rate (measured by the retail price index RPI) minus the sectoral productivity growth rate. This is basically the approach adopted by the EU and by many other countries.

As regards mobile telecommunications, government is involved in the allocation of spectrum which is a scarce resource. A critical question involves the size of the relevant market and digital international market integration. If large international markets are relevant, one should expect the formation of large mobile telecommunication companies which in turn could be facilitated by secondary markets for mobile licenses. Such secondary markets are non-existent due to poor rules set by government in the context of allocation of mobile licences. Moreover, as regards the EU there are distortions of services and capital markets through the divergent ways of allocating spectrum. While auctioning seems to be the appropriate way for efficient allocation, several countries have combined a beauty contest with a modest fee. A critical issue in mobile competition involves roaming fees which in some EU countries seem to be higher than long run marginal roaming costs.

Mobile telecommunications is a leading field of digital innovation dynamics, including the development of novel services. As regards substitutability of fixed network communications and mobile telecommunications, it is clear that there is only weak substitutability. This holds despite the fact that among young users, mobile telecommunications is often the first choice when it comes to getting their first own number. This holds particularly in eastern Europe. However, for certain services – including internet services – fixed network access is much better in terms of speed and certain convenience aspects than mobile access.

Within less than a decade, mobile density has exceeded the fixed line density in the EU and Japan; only the US still records a relatively lower mobile line density; however, the average revenue per user in the US exceeded that in the EU in 2005 so that prospects for US catching up are favorable. Since mobile phone calls are more expensive than fixed line calls, it is clear that mobile sales – and (probably) their respective profits – exceed that of fixed line sales. Thus firms operating both in the fixed line business and in the mobile business have an incentive to further push the mobile business. Technological progress which gives mobile users a so-called cheap home zone – this amounts to using the mobile phone at home in a similar way as a traditional fixed line phone – encourages users to replace fixed-line telephony with mobile telephony. In 2004/05 internet-based telephony (VOIP) has become a new alternative for millions of households worldwide, and there also is an increasing number of business users which will rely on VOIP which economically eliminates the distinction between local calls and international calls. Major international organizations such as the EU, the World Bank, the WTO etc. could save a considerable share of the respective budget if they would switch to VOIP, and there is no reason why not all UN organizations should switch to this technology at the same time.

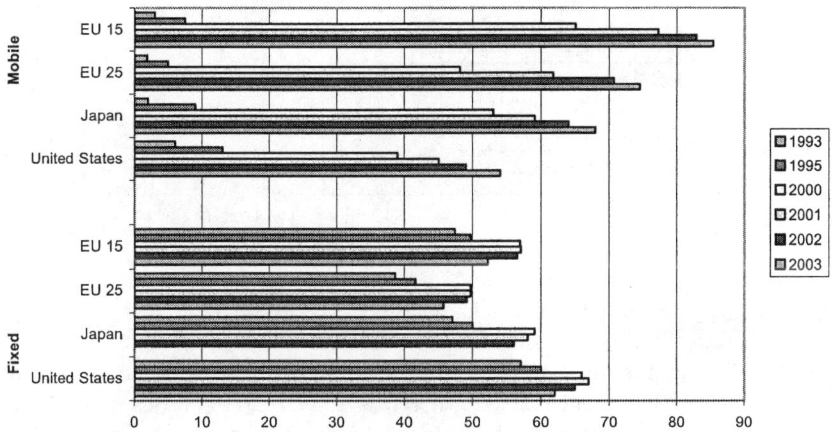

Source: ITU (2004)

Fig. 2. Mobile and fixed network density in selected countries

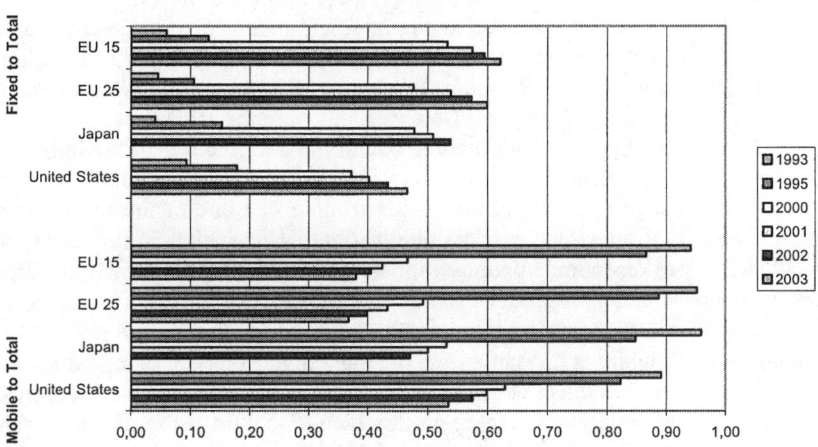

Source: ITU (2004)

Fig. 3. Total share of mobile and fixed networks in selected countries

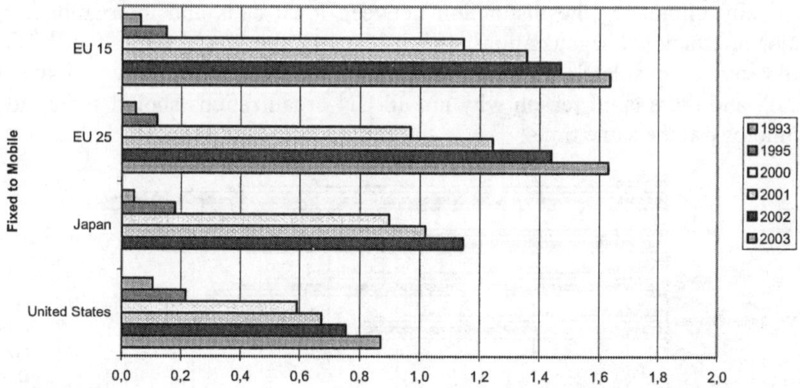

Source: ITU (2004)

Fig. 4. Ratio of mobile density to fixed line density in selected countries

As regards the internet there was an enormous growth of both the supply side (host density) and the demand side (user density) in OECD countries in the 1990s. By 2010, about 70% of households could have access to the internet. Governments in most OECD countries have actively encouraged the use of the internet through both special programmes to link the schools to the internet and via the provision of e-government services. It is, however, not clear whether governments in all countries worldwide are interested in stimulating the use of the internet.

As regards the use of PCs as well as the internet, there is a strong North-South division in the world economy as is shown in the subsequent table. As regards relative prices, Japan, Korea, Belgium, Hongkong, Singapore, the United States, Canada, the Netherlands, Macao and New Zealand were the 10 cheapest countries in 2003. Finding the US as a high income country among the top 10 could be explained by cross-subsidization of such services or by powerful network effects or rapid technological progress. It is interesting to note that one of the most costly countries is Saudi-Arabia. One may assume that in this case, high pricing is a government strategy to discourage people from using the global internet with its large diversity of views and independent source of information. A particular problem with the digital North-South gap concerns Africa which has the world highest population growth while at the same time having a high illiteracy ratio (except for South Africa and a few select countries). Thus the politically popular perception that a digital world economy is emerging in which global diffusion of knowledge contributes to economic and technological catching-up is partly wishful thinking.

Table 1. Internet prices (broadband) in selected countries

Nr	Economy	Subscription per month (USD)	Prico per 100 kbit/s (USD)	Subsription as % of monthly income	100 kbit/s as % of monthly income
1	Japan	24.19	0.09	1.11	<0.01
2	Korea (Rep. of)	49.23	0.25	3.58	0.02
3	Belgium	34.41	1.15	1.51	0.05
4	Hong Kong, China	38.21	1.27	1.71	0.06
5	Singapore	33.18	2.21	1.69	0.11
6	United States	52.99	3.53	1.81	0.12
7	Canada	32.48	3.25	1.39	0.14
8	Netherlands	51.55	3.36	2.25	0.15
9	Macao, China	38.34	2.56	2.43	0.16
10	New Zealand	40.61	2.71	2.43	0.16
11	Germany	33.93	4.42	1.55	0.20
12	Norway	46.16	6.56	1.55	0.22
13	Israel	20.40	3.98	1.27	0.25
14	Austria	45.20	5.89	1.92	0.25
15	Slovenia	79.54	3.88	5.40	0.26
16	Italy	73.59	6.13	3.49	0.29
17	United Kingdom	32.59	6.37	1.51	0.30
18	Luxembourg	91.77	17.92	2.16	0.42
19	Sweden	44.56	8.91	2.13	0.43
20	Switzerland	57.84	11.30	2.22	0.43
21	Australia	50.56	9.87	2.25	0.44
22	France	51.46	10.05	2.36	0.46
23	Ireland	61.69	12.05	2.64	0.52
24	Portugal	39.64	7.74	2.74	0.54
25	Cyprus	58.03	9.07	3.86	0.60
26	Iceland	73.66	14.39	3.09	0.60
27	Lithuania	12.80	5.00	1.55	0.61
28	Malta	53.34	10.42	3.77	0.74
29	Jordan	14.06	2.75	4.15	0.81
30	Denmark	51.82	20.24	2.11	0.82
31	China	30.10	7.84	3.70	0.96
32	Croatia	24.26	9.48	2.62	1.02
33	Estonia	49.72	4.86	10.58	1.03

34	Venezuela	42.95	11.18	4.02	1.05
35	Hungary	57.36	22.41	2.71	1.06
36	Finland	47.63	18.61	2.79	1.09
37	Spain	29.21	7.61	4.23	1.10
38	Malaysia	68.90	13.46	7.03	1.37
39	Argentina	22.44	8.77	3.71	1.45
40	Brazil	71.19	27.81	3.89	1.52
41	New Caledonia	76.15	14.87	9.02	1.76
42	Poland	35.50	13.87	4.64	1.81
43	Chile	106.10	41.44	8.01	3.13
44	Bahrain	57.46	22.44	8.01	3.13
45	Mexico	75.31	29.42	10.11	3.95
46	Latvia	80.00	31.25	11.62	4.54
47	Costa Rica	99.00	19.34	24.75	4.83
48	Peru	93.26	36.43	16.58	6.48
49	Grenada	238.65	93.22	24.65	9.63
50	Saudi Arabia	174.75	68.26	31.39	12.26

Source: ITU, World Telecommunication Development Report 2003

2.4 Innovation Aspects

As the share of innovation expenditures in ICT rises relative to R&D expenditures in other sectors (OECD, 1999), the role of "digital innovations" will increase. This could hold more so with respect to the R&D input side than with respect to R&D output (here we have an empirical question). One the fascinating features of the internet is that it allows more services to be traded as providers of the respective service or the user effectively becomes more mobile in a geographical sense. This should stimulate competition and hence efficiency gains and innovations. Leading digital OECD countries can thus be expected to have a high share of product innovations in services.

Product innovations allow for the increase in product prices in world markets and hence the earning of high incomes (wages and profit). With digital products and services market through the internet, the opportunities for increasing prices for novel goods is somewhat restricted through the global nature of the internet and the rather limited opportunities for price differentiations. Digital products could, of course, be presented as differentiated products to some extent as the example of the differentiated music CD launched in 2004 revealed. Certain services can individually be customized by the user himself, for example in the field of news surveys.

Process innovations are equivalent to cost reductions and allow firms to fetch higher market shares and high incomes, in particular if price elasticity is larger than unity or if increased market share also allows for the exploitation of dynamic scale economies (e.g., learning by doing effects). Innovation dynamics can be assessed in different ways:

- Innovation expenditures, usually scaled by sales ("R&D intensity"); this in an R&D input indicator
- Patents per capita (R&D output indicator)
- Product innovation rate (new products to the market in % of sales, survey data, innovation output indicator); ICT facilitates product innovations and should hence stimulate intra-industrial trade. Internet based services innovations might be more difficult to combine with price differentiation
- Diffusion rate (new-to-the-firm products; figures are from surveys): the diffusion rate could accelerate in the digital society

Taking a closer look at selected EU countries as well as the US and Japan, one finds that Sweden, Germany and Finland were leading in R&D intensity in manufacturing (6.4, 4.7 and 3.9, respectively, in 2003; EU average 3.45; see Tab. 1). France and the Netherlands achieved 3.1, the UK 3.0. Germany's R&D intensity in the services sector was much weaker, namely 1.6 compared to the EU average of 1.8. Sweden was a clear leader in this field. France and the UK recorded 1.6 and 1.4, respectively. It is interesting to observe that in the field of product innovations in manufacturing, Germany was below the EU average despite its leading position in R&D intensity. Finland, Sweden and France were leading countries in the field of product innovations. This suggests that the German innovation system might have considerable efficiency problems. A similar picture is found in production innovation in the services market. As regards diffusion indicators, Germany is a leading EU country. Moreover, Sweden and Germany recorded a high ratio of New-to-firm to New-to-market in manufacturing industry, which points to relatively fast diffusion (this could reflect strong competition). It is not known to which extent the strong role of Scandinavian countries in services innovations is linked to their strong position in ICT.

Table 2. European innovation scoreboard, 2003

European Innovation Scoreboard 2003 - Member States, US and Japan

	EU 15	DE	FR	NL	AT	FI	SE	UK
Innov exp manuf	3.45	4.71	3.08	3.07	2.83	3.91	6.42	2.96
Innov exp serv	1.83	1.64	1.57	0.79	0.92	0.96	19.11	1.39
New-to-mark prods manuf	10.5	7.1	9.5	-	8.4	27.2	3.5	9.5
New-to-mark prods serv	7.4	3.7	5.5	-	4.3	12.2	9.3	-
New-to-firm prods manuf	28.6	40.3	17.5	23.8	23.1	31.1	32.1	-
New-to-firm prods serv	18.8	16.4	17.1	13.9	12.8	18.8	23.7	-
New-to-firm/New-to-mark prods manuf	2.7	5.7	1.8	-	2.8	1.1	9.2	-
New-to-firm/New-to mark prods serv	2.5	4.4	3.1	-	3.0	1.5	2.5	-

Source: European Commission (2003), Staff Working Papers, European Innovation Scoreboard 2003, page 27, Brussels and own calculations.

The careful exploitation of opportunities to raise productivity in the information and communication technology (ICT) could be a new and important policy element (BARFIELD/HEIDUK/WELFENS, 2002). Raising labour productivity was an important element of high growth in the US and several EU countries in the 1990s. While there is no debate about the productivity-enhancing role of ICT production, it is less clear whether the use of ICT – linked to ICT capital accumulation – strongly contributes to higher growth of output and productivity. (A strange case is productivity measurement in the US retail sector where a firm with 10 employees selling 10 standard PCs in 1999, but 10 more modern PCs in 2000 – worth 5 times as much according to hedonic price measurement as a 1999 PC – shows productivity growth of 500%!). Production of computers or telecommunication equipment seems to be crucial for growth.

Given the increasingly important role of innovation dynamics for international markets, the promotion of research and development becomes a crucial part of economic policy. The traditional argument in favour of R&D promotion is the existence of positive external effects which imply that marginal social benefit exceeds marginal private benefits of R&D expenditures. However, some new developments in innovation dynamics have to be taken into account when raising the issue as to which role government should assume in the promotion of innovation and skills. In a networked society there will be many markets with network effects (endogenous growth of demand), which in certain sectors will also go along with static and dynamic scale effects. This implies that there will be large differences between monopoly outcome and a competitive market solution. If there is product innovation in a market with network effects and scale effects there could be a sustained monopoly which would imply large welfare losses. The situation is less worrisome if the innovator has only a temporary monopoly while newcomers bring about a competitive solution in the long run (see appendix).

2.5 Digital Media Society

In the digital social market economy, there will be a strong stimulus for the expansion of all typres of modern (diverified) TV services (WELFENS/ZOCHE/JUNGMITTAG/BECKERT/JOISTEN, 2004). Technological convergence which blurs traditional market demarcations for transmission of voice, video and data is a key element behind this. TV programmes can be transmitted via satellite, cable TV networks or telephone networks, including mobile telephony.

The standard way to organize TV is peculiar and does not resemble a normal market. Countries with state-owned TV stations finance the cost of programmes through monthly fees plus revenue from advertisement; public TV programmes are considered to be part of universal services which government wants to offer. Many countries also have private TV operators which finance programmes exclusively via advertisement (and TV gambling based on call-in games). Pay-TV would be the normal market solution for TV information and entertainment, but it plays a remarkably small role. It can be argued that the dominant form of private "free" TV stations relying on advertisement has negative welfare effects and thus is a doubtful exercise. It has many elements of a socialist system in the sense that the viewer gets a free lunch while apparently not having to pay for this. However, it is clear that those who buy the goods advertised – which will have higher prices due to TV advertisement costs – will ultimately pay for free TV.

The problem is shown in the following graph where the TV market is shown in the left-hand panel and the market for soft drinks (cola) in the right-hand panel. In the TV market, we have falling average costs. Advertisement reduces the demand for free TV, because there are opportunity costs of time. The demand is DD^*_1 instead of DD^*_0. Consumers view the output Q_0, where Q_0F^* is the opportunity cost of viewing advertisement. In the soft drink market, the marginal costs curve is shifted upwards as TV advertisement requires to pay for advertisement time in many channels (the most popular channels represents the lowest marginal costs of advertisement, the least popular channel the highest marginal costs – assuming no income differences across viewers of different channels). Hence there is an upward shift of the marginal costs curve; at the same time the demand curve will shift upwards due to the advertisement effect. The equilibrium output falls from qo to q1 while the price increases from p_0 to p_1. Advertisement revenue is equal to the area $p_1E_1CC^*$; it must be large enough to cover TV costs which are Q_0D^*DO.

The combined welfare effect of both markets is unclear but it could be negative. Here empirical research is required.

The indirect way to finance free TV programmes lets us expect that the quality is rather weak and mainly determined by the marketing bosses of firms which produce products with low elasticity of demand – advertisement for such goods let us expect high additional revenues. Hence viewer preferences are only partly reflected. Worse, since marketing rules say that only viewers in the age bracket 15-45 are relevant, the preferences of the majority of the population (in an ageing society) might be fully ignored. This is in sharp contrast to consumer sovereignty, which is a standard hallmark of a functional market economy.

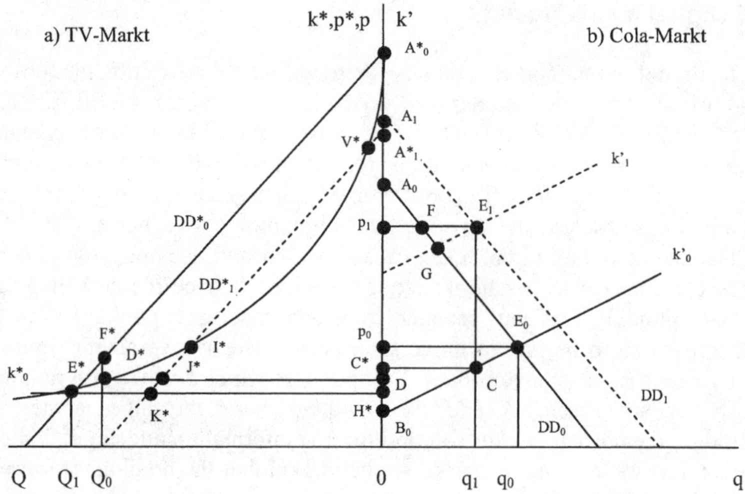

(welfare loss in TV-market ≈ $A^*_0V^*I^*E^*+Q_0D^*I^*$; welfare gain $= J^*D^*OQ_0$; welfare loss in cola market $= P_1FE_0B_0$; welfare gain $= A_1E_1GA_0$)

Fig. 5. Advertisement as TV-financing: negative welfare effects

2.6 The Software Economy

Software expenditures relative to GDP accounted for about 2% of GDP in leading OECD countries at the end of the 20th century and thus come close to the size of R&D expenditures. Interestingly, the number of software engineers exceeded the number of all engineers employed in conventional fields in 2000 in the US, which shows there is a hard core of the New Economy. Software allows flexible automatization, controlling and providing digital services in almost all sectors. As workflows must be analyzed as a first step before programming, there is automatically a learning process with respect to simplifying and standardization. Moreover, workflow can be decomposed into subelements which in turn can be recombined for innovative future applications.

Software markets at the beginning of the 20th century can be characterized in several ways:

- rapid growth;
- strong innovation dynamics;
- ongoing internationalization
- ongoing international outsourcing
- mixture of consolidation in some fields and rise of small newcomers in new fields
- rising complexity which raises the need for more powerful computers and chips, respectively.

Compared to hardware, the relative price of software is increasing in the long run, which seems to be due to a combination of static and dynamic scale economies (plus Moore's Law for technological progress in chip design) in the hardware business and increasing demand – plus rising market power – in the software market. To the extent that a problem can be solved alternatively through hardware or software, it makes sense to prefer the software solution since it is more flexible and allows easy updating. Market power of certain software firms certainly is a crucial problem in the digital social market economy, since the endogenous monopolization dynamics of certain market segments could lead to high monopolistic pricing, as has become obvious in the Microsoft case. The US decision to introduce software patents is particularly doubtful with respect to those markets where network effects are strong. The network effect itself is a mechanism which favours first-mover advantages and the firm with the largest market share. Software patents therefore should, if introduced at all, have a much shorter duration than the standard 20 years relevant in patent laws designed for an industrial world.

Software design can only partly be outsourced internationally. Complex software always has to be tested in close collaboration with the lead users which typically are located in leading OECD countries. From this perspective, the (advanced) software sector to a certain extent can be considered as an "immobile Schumpeterian industry". It is technology-intensive and thus dubbed Schumpeterian, and it is immobile in the sense that the whole sector cannot be relocated from high wage OECD countries to low wage NICs or developing countries.

Reliability of software is worrisome to some degree. This holds less for advanced software used in a chemical plant or by the military, for instance. However, software used in the context of the internet (and by households on PCs) tends to have a relatively high failure rate. This is troublesome in a world in which internet activities are expanding more and more in every country and in which the internet is a truly global platform, so that negative international spillovers can occur. Virus attacks on servers are a well-known problem, and it is rather unclear whether users of computers and the internet take into account the potential negative external effects which insufficient virus-control of digital products really has. From this perspective, product standards in digital security are insufficient. As long as governments fail to cooperate worldwide, this problem is likely to continue.

2.7 Competition, Efficiency and Welfare-maximizing Differentiated Two-Part Tariffs

From an economist's perspective interested in sustained competition, one should not impose regulations without clear reasons. The EU which has imposed a broad regulatory framework – and holds a double veto power (with respect to market definition and market analysis as provided in the notification by the national regulatory authorities) – requires a triple cumulative test:

- There must be considerable sustaining barriers to entry;
- there is no inherent tendency in the market to achieve effective competition within a relevant time horizon
- general competition law is not sufficient to provide an adequate remedy for the problems to be anticipated.

There is, however, no common degree of regulation across EU countries. Even if one takes into account that initial conditions differ across countries, it is rather surprising to see that the ECTA Regulatory Scorecard (a study sponsored by OFCOM) shows large differences across EU countries, the leading countries being the UK, Denmark and France. Poland, Greece and Germany are positioned at the bottom of the rankings. ECTA has argued that there is a positive correlation between the scorecard results and the level of investment in the telecommunications sector.

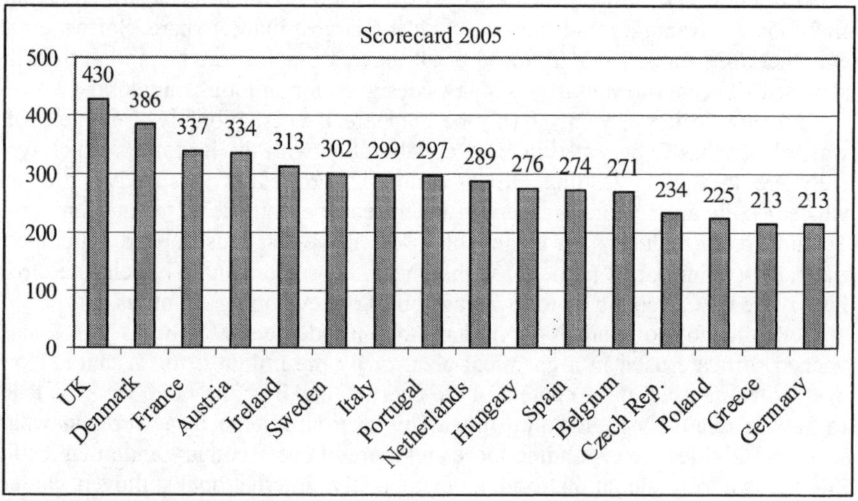

Source: ECTA (2005), Regulatory Scoreboard.

Fig. 6. Regulatory scorecard

To some extent, foreign direct investment should contribute to a narrowing of the scores across countries. There is certainly some regulatory arbitrage across EU countries, at the same time one cannot overlook that large EU countries – representing relatively important markets – still enjoy considerable policy autonomy. Whether a stricter EU regulatory framework would be useful is doubtful since it is unclear whether or not the EU will pursue a rational policy approach which would enhance static and dynamic efficiency. It is quite interesting to notice that there is a clear positive correlation between the scorecard results and investment in the telecommunications sector (ECTA, 2005). As this sector is part of the ICT sector with its high technological dynamics, low investment growth in the telecommunications sector is likely to undermine overall economic growth. From a theoretical and empirical point, there are also other arguments which point to a particular relevance of the telecommunications sector in a digitally-networked economy. For facilities-based competition, this is obviously crucial.

Among the thorny regulatory policy problems is the relationship between general competition law and regulations. Only a consistent framework at the national level will optimally stimulate static and dynamic efficiency. OFTEL/OFCOM has enjoyed both regulatory powers (ex ante interventions) and a role in the broader field of competition policy which mainly is ex post intervention. In Canada, the Competition Bureau and the Canadian Radio-Television and Telecommunications Commission (CRTC) have concluded an Interface Agreement in 1999 specifying those areas where each authority has jurisdiction and fields of joint responsibility. Similarly, there was an agreement – the Co-operation Protocol – between OPTA, the regulator in the Netherlands and the Dutch competition authority (OECD, 2006). The revised EU Framework Directives, which emphasizes market analysis (basically for 18 markets) before implementing regulations, requires that member states ensure that "this analysis (market analysis) is carried out, where appropriate, in collaboration with the national competition authorities". As regards the US, authorities avoid potential conflicts associated with overlapping jurisdiction over competition issues. The FCC has concurrent authority with the Department of Justice to enforce section 7 of the Clayton Act with respect to telecommunications common carriers. Moreover, as regards mergers the FCC does not need to rely on the Clayton Act Section 7, rather it can rely on its more general "public interest" authority to review transfers of licenses or authorizations subject to regulations of FCC. In merger cases, consistent decisions require cooperation between the FCC and antitrust agencies which can be difficult in the case that confidential information available to the latter cannot easily be shared with the FCC (a waiver often helps here). There is no formal cooperation agreement between the FCC and the DOJ and the Federal Trade Commission, where informal cooperation seems to work. The FCC also often refers to merger guidelines jointly developed by the Federal Trade Commission and the DOJ for standard principles on issues as market definition and measurement of concentration. However, one may pose the question as to whether in an increasingly internationalized digital market and with rapid growth of VOIP/internet-based telephony, will there be a global telecommunications market in which it might become difficult to distinguish between EU market and US markets. Enforcement in the internet world is difficult.

A growing share of digital services provided through broadband is related to internet providers. For internet providers access to high speed networks is crucial.

Here vertical integration on the side of the incumbent operator can cause conflicts. If the incumbent has a dominant market position and wants to exclusively provide high speed services to customers this raises the question about undiscriminatory access. Taking a look at US developments and EU developments in 2004/05 it is not clear that one can easily find a consistent regulatory answer for this challenge – considering the transatlantic regulatory dialogue is seems that one might not find a common approach on both sides of the atlantic. Whether or not the OECD is a policy forum developing guidelines on digital access of internet providers remains to be seen. As regards internet providers interested in launching innovative services there is a natural interest to obtain bitstream access and to offer a range of various services packages which might include VOIP telephony whose expansion could cause accelerated depreciation on the side of the incumbent fixed-line operator; the stock market price also could come under pressure.

Strong competition in EU markets – after opening up in 1998 and a wave of international investment which often reflected a mutual invasion of neighbor markets through former national monopoly operators – has strongly reduced national and international telecommunications prices. The liberalization of markets worldwide has eroded the role of the International Telecommunications Union which traditionally has set accounting and settlement rates, that is, organized the revenue split between outgoing traffic countries and inbound traffic countries. With asymmetric liberalization in the 1980s and early 1990s it was the US which had the highest sectoral current account deficit in telecommunications services since relatively high prices abroad implied a bias for international calls to originate in the US.

A relative fall of international telecommunications prices will stimulate internationalization of digital services as will strong competition in local access markets, in particular in DSL and other broadband segments; cheaper local access facilitates the expansion of internet service providers. The internet plays a key role in the World Summit on the Information Society (WSIS) which is long run project under the auspices of ITU.

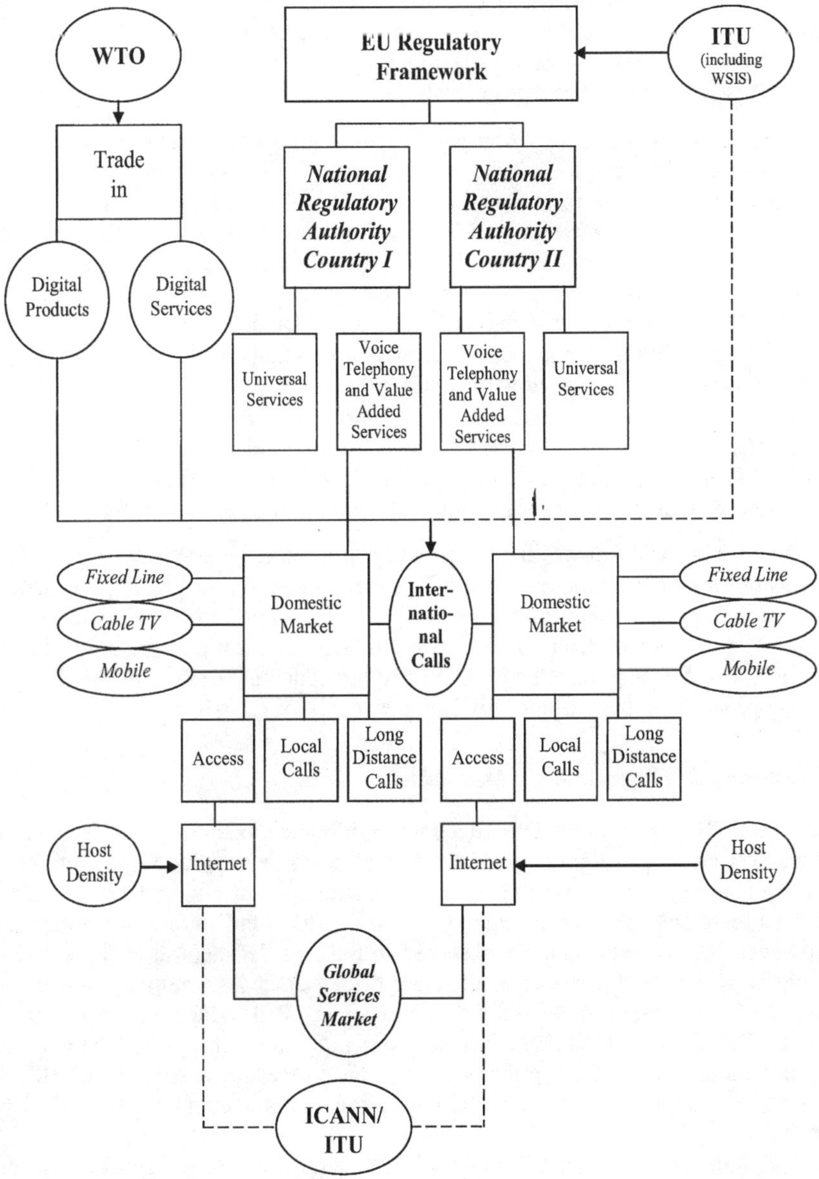

Fig. 7. Regulation of telecommunications and digital services

Certainly, in all EU countries there is long term digital growth. Moreover, international outsourcing takes place – partly reflecting digital outsourcing -, rising trade in ICT goods is observed and e-government has become prolific. There are considerable differences among EU countries/OECD countries, respectively. This is partly related to two basic problems:

- the fact that (following the arguments of ARROW) information markets are systematically imperfect as those who want to sell digital information have to reveal part of the information for free. This implies that national information markets as well as international information markets typically suffer from market imperfections related to the fact that the effective demand – as perceived by the supply side – is below the demand level reflecting the marginal social benefit. This also implies that the national excess demand in any potential import market for information is lower than in a normal market and hence international trade in information is below the optimum level.
- Part of the information and knowledge market suffers from the fact that the respective products are "experience goods" or "confidence goods" where potential users find it quite difficult to indicate the value of the respective goods. Hence there is a market for lemons problem in part of digital markets.

While it is well known that in markets with experience goods or confidence goods building reputation on the side of sellers and certain signaling mechanisms can minimize market failure, the real digital world is not always characterized by efficient strategies of firms. International telecommunications firms and internet service providers also face the problem that they encounter rather different regulatory approaches in the various OECD countries.

Efficiency in Oligopolistic Markets

In several OECD countries liberalized telecommunications markets effectively are characterized by an oligopoly or the tendency towards an oligopoly. There are all kinds of models for various types of firms' strategies in oligopolistic markets so that there seems to be broad uncertainty about the implications for regulations. However, the typical oligopoly observed in the first decade(s) after liberalization is one in which the incumbent is the clear price leader. Assuming asymmetric behavior of newcomers in the sense that newcomers will reduce the price if the incumbent is reducing the price while newcomers do not follow when there is a rise of the incumbent's offer price we have an interesting case of the HITCH-SWEEZY model, in a setup with falling marginal and average costs (WELFENS, 2004).

The assumptions of the HITCH-SWEEZY model are rather specific and only empirical analysis can tell whether the reality in fixed-line telephony indeed is represented by the assumed type of asymmetric behavior – a variant of "follow-the-leader" strategy. In the model there is an asymmetrical interdependency under oligopoly as there is one dominant firm and many relatively small competitors: If there is price reduction of the dominant supplier 1 the other firms will follow, which makes part of the effective demand curve less price elastic (steeper than a normal demand curve); the arch below the current equilibrium price is rather inelastic. If firm 1 raises the price other firms will, by assumption, not follow so that the effective demand curve is more elastic above point B: see the segment BA‘. If the oligopoly is facing less competition through substitute products – e.g. as a consequence of bundling of products in the market – the demand curve becomes steeper above point B: see segment BA; temporary pricing according to marginal costs implies a reduced quantity in this situation with a more pronounced price leadership (compare point F’ and F). If the situation is a stable oligopoly with clear leadership there is not a big difference between the hypothetical optimum defined by equality of marginal costs and the demand curve and pricing based on average costs (point B and equilibrium quantity q_0 versus optimum output q_1 and point E, respectively). The problem of a dominant market position – and hence the situation of a leading leader - is likely to emerge in an oligopoly in which one firm has a clear lead in the market share. The existence of network effects would not really alter the argumentation, only the equilibrium point is moving from E (or B) to E’. The key aspect in a Hitch-Sweezy model with falling marginal and average costs is that the difference between the optimum allocation q_1 and q_0 is rather small; and given the costs of regulations one might then indeed consider only relatively limited regulation at all as long as the asymmetric oligopoly is persistent. The situation is more complex if one considers vertical integration which potentially raises problem of cross-subsidization and discriminatory access on the side of competing service providers.

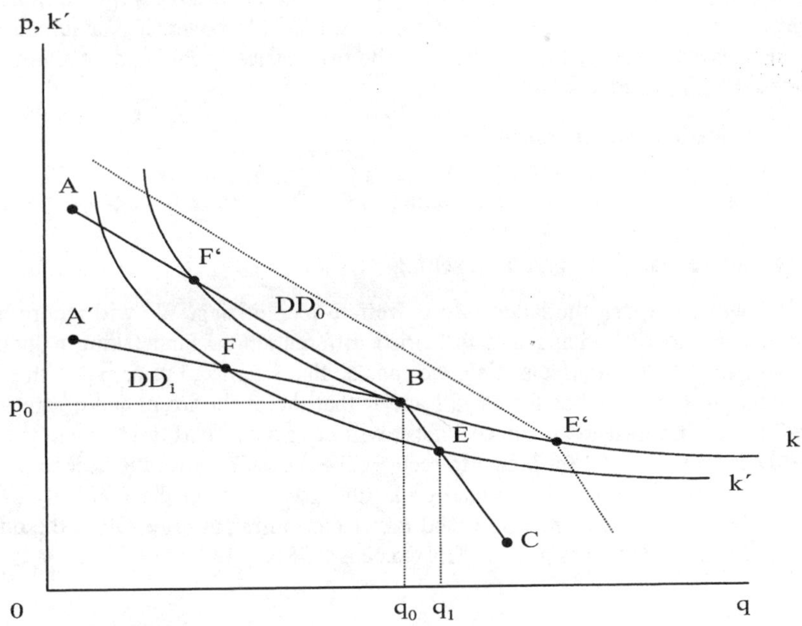

Fig. 8. Modified Hitch-Sweezy approach to oligopoly

Achieving an Optimum through Differentiated Two-part Tariffs

Two-part tariffs are rather common in network industries. However, to date there is no clear theoretical model. The following reflections present a brief argument not just in favor of two-part tariffs but of differentiated two-part tariffs which could be highly relevant for regulators in telecommunications. Indeed, one could argue that uniform regulated prices at the wholesale level or in the access market are not Pareto-efficient, rather differentiated two-part tariffs are required. If the dominant operator is not offering differentiated two-part tariffs the regulator should impose a provision which enforces broadly differentiated two-part tariffs. The following arguments should make clear the logic behind this approach.

Two part tariffs are used in many sectors of the economy, in particular in network industries where high fixed costs are important. The firm can recover costs more easily if it does not offer a uniform price but a two part tariff which is composed of an access fee and the option for the user to buy a certain quantity at marginal costs. If the marginal costs are constant the arrangement almost looks like a uniform price except that it is composed of two elements, of which one is the access fee and the other the uniform price. KNIEPS (1998) has argued that such a two part tariff is welfare enhancing. We will refine the argument in a broad way and argue that differentiated two part tariffs will bring enormous welfare gains. At first we will take a look at the simple two part tariff, namely for the case of two users with demand DD_I and DD_{II} respectively. This corresponds to the case of KNIPES (1998). Aggregate demand is given by the curve DD_{I+II}. For simplicity, marginal costs k' are assumed to be constant, k'_0 . Moreover, it is assumed that fixed costs are given by the area PTUN. The firm basically could adopt alternative price strategies as follows:

- (i) standard monopoly pricing
- (ii) full price differentiation
- (iii) cost-oriented uniform price setting
- (iv) two-part tariff setting
- (v) differentiated two part tariff setting

We will show that the latter case is welfare maximizing. We will ignore here standard monopoly pricing and full price differentiation. First taking a look at cost-oriented uniform price setting, we can see that a uniform price p – with p being equal to OP – implies for users I and II that the quantities q^I and q^{II} are realized. If the firm instead announces that it will offer a two part tariff where paying the access price PTUN (designed to recover fixed costs) allows the user to obtain his/her desired quantity at marginal costs, the user I – with the relatively small demand – will renounce this offer and rather consuming at price OP=p the quantity q_I. Clearly, the access price PTUN exceeds PQSN.

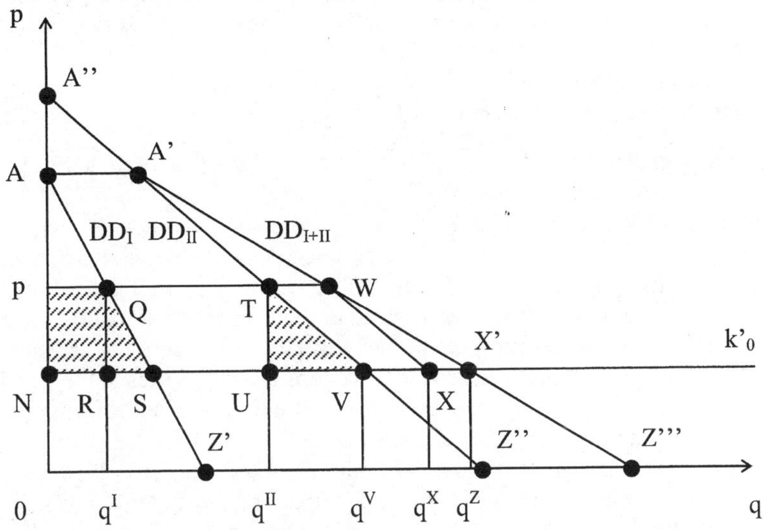

Fig. 9. Output and welfare effect of introducing a differentiated two part tariff

The high volume user II will, however, prefer the two-part tariff so that he/she will have a demand equal to q^V, which exceeds q^{II}, the overall demand thus increasing to q^X. The welfare gain from switching to the two part tariff is given by the triangle TVU.

We now can go beyond the traditional approach. It can be argued that differentiated access cards/access prices are welfare maximizing. The firm offers individual access cards to every user. For the user I the adequate access price is PQRN – demand of user I will be equal to the distance NS. Offering indeed two different access cards thus raises the equilibrium demand by the distance RS. User II gets an access card at the price QTUR. With the differentiated access card, we now have a further rise in the equilibrium quantity which now is equal to q^Z. The welfare gain from switching to a differentiated two part tariff is the triangle QRS plus the area PQRN.

We immediately recognize that the standard procedure by certain regulatory authorities of imposing uniform access charges is not welfare-maximizing. Instead differentiated two-part tariffs are welfare maximizing (also at the wholesale level). They achieve the same goal as Ramsey pricing, but in a different way. In our approach it is not the regulator which has to estimate cross price elasticity's, as in the case of Ramsey pricing. Rather the regulator would stipulate that users are entitled to individual entry cards on the basis of an efficient provision of services. Here this means that the regulated firm should not be allowed to inflate its fixed costs, thereby reducing marginal user prices artificially. The regulator only needs knowledge about the cost structure of the regulated firm; and the welfare gain from switching to a differentiated two-part tariff can be high (see appendix). An important implication is that equal access requirement cannot simply be translated into equal prices for all network users; in the special sense of a differentiated two-part-tariff discounts should be allowed.

As regards welfare analysis in regulatory policy one should emphasize that a comprehensive analysis should consider:

- Welfare analysis in the sense of consumer surplus for households in the telecommunications services sector.
- Welfare analysis in the sense of taking into account the producers surplus both in the domestic economy and in foreign countries – the latter to the extent that the domestic regulatory framework affects the ability of multinational telecommunications firms to be successful foreign investors abroad (from a pure national policymaker's perspective one normally will ignore the producers surplus in the home market which accrues to foreign subsidiaries in the country; in the EU context, however, the European Commission in its regulatory policy should consider overall EU25 welfare analysis).
- Welfare analysis should take into not only the direct impact in the telecommunications services market but also the important direct effects which occur through telecommunications services used as intermediate inputs in all sectors of the economy: E.g. a fall of telecommunications services prices will translate into lower costs of production and hence higher volumes sold of all products using telecommunications services. As the volume of telecommunications services for business customers in OECD is higher than services sold directly to households one should indeed not overlook this aspect. A similar reasoning applies to product innovations supported by innovative digital services inputs (upward rotation of the demand curve).

Finally, with international network effects in telecommunications the traditional view of national policymakers also becomes less convincing. The stronger foreign direct investment is in telecommunications and the more relevant international network effects are the more is there a need for a broader international policy perspective and for policy cooperation.

3 New System Design and Strategic Policy Aspects

Sustained competition in telecommunications markets – in particular in fixed-line telephony – is viable only with minimum ex-ante regulation. Without a regulator in natural monopoly sectors, interconnection will be inefficient and problems of predatory bundling will remain unsolved. Since first-mover advantages in dynamic telecommunication markets and the problem of market power are important in fixed line telecommunications, it is clear that there should be regulation. Excessive regulation, however, should be avoided. Price-cap models have turned out to be a useful approach to reconciling the goal of efficiency and innovation. A crucial issue is to determine the relevant market, which is not easy in the context of technological convergence and progress. With ongoing globalization and regional integration in all continents – and taking into account the impact of a long term fall in international telecommunication prices – there is a long term tendency for larger markets to emerge. Thus regulation will face internationalization challenges.

The internationalization of goods and services markets will partly be driven by both internet and e-commerce. Here the WTO must come up with adequate rules for digital products (falling under GATT) and digital services (falling under the weaker set of rules established within GATS). A rather difficult question concerns the problem of overshooting, which could be considered a natural topic for the Bank of International Settlements which is a kind of watchdog for the stability of global financial markets. Joint international initiatives on the reform of national tax codes in a way that more long term investment is encouraged could be one element of a future policy (which could more realistically involve the OECD than the BIS, since the OECD has a focus on some taxation issues including internet and taxation).

A complex problem related to static and dynamic efficiency in the digital economy concerns adequate intellectual property rights. As regards software there is no need to introduce standard duration of patents; 5-10 years should suffice since network effects create endogenous market entry barriers. One might consider waiving software patents immediately once the product has a market share of more than 50%. More generally, digital copyrights are relatively weak which could undermine the incentive to develop digital innovations and to produce quality information for a dynamic knowledge society. One example of setting digital intellectual property rights can be seen in Germany where government translated the EU e-commerce directive into national law in 2003 in a rather strange way. The law allows digital communities – say all university libraries – to cheaply tap into print media whose content is also available in digital form. At the same time, politicians want most research published free-of-charge in the internet. Those writing quality text for publication and those launching innovations have almost no incentive; rather government encourages intellectual property rights to be violated, and scientific publishers will be encouraged to move out of Germany. If digital property rights should be weaker than those for ordinary bricks, the way back to the stone age is wide open. (Switzerland and the Netherlands had abolished national patents law in 1849 and 1860, respectively, because politicians thought that no patent protection is the best way to promote innovations. They discovered, however, that this drives innovators out of the country and reduces innovation output sharply, since conducting R&D – a risky business – is no longer really rewarded with prospects for a premium income.)

As regards e-commerce, unified forms in public tendering would save enormous transactions costs. It is surprising to see the bewildering variety of forms used by government authorities. Efficiency gains and learning effects will be relatively large if standardization is realized where necessary and useful. A serious problem specifically related to the internet is internet security including secure payments. It is difficult to understand why credit cards developed for a non-digital world are popular in internet transactions. The G-8 should launch a common project to develop a novel means of payments which is secure for internet transactions. This would boost the volume of transactions on the net.

The digital economy will clearly require standard setting by industry itself. Open standards certainly are quite useful. It remains to be seen whether the ICT subsectors will be characterized by efficient self-regulation in the sense that industry standards are established by international consortia of leading firms. The recent example of Phillips – with its new Science Park in Eindoven in which Phillips researchers cooperate with several hundred firms also conducting research and swapping licences among each other – shows a new approach to global standard setting through networked innovation efforts guided by a global leader.

In the digital social market economy equal, opportunities are no less crucial than in a traditional market economy. Avoiding a digital divide in a society is therefore quite important as is the issue of avoiding a North-South divide in the context of a global economy. Equal opportunity could mean that governments would embrace a broader defined universal service. An EU-wide or global broadband universal service which could include internet telephony has been proposed (WELFENS, 2004b). Competitive international tendering would allow provision of such new services at minimal costs. An OECD country could devote part of traditional development policy funds to help achieve the new goal for poor countries.

A rather inefficient sector not compatible with a functional market economy is the TV market in which so-called free TV of private providers could create considerable negative welfare effects. A tax on TV advertisement would be an adequate way to internalize negative external effects from free TV.

As regards global networking and optimal regulation of telecommunications services, ITU could play an important role. The International Telecommunications Union no longer fixes accounting and settlement rates for international telephony since this task is largely obsolete in the new competitive global environment. However, ITU could have an important role to help countries – particularly poor countries – to adopt adequate regulation for fixed line telephony and for mobile telecommunications. To the extent that adequate pro-competitive regulation reduces prices and stimulates innovation, the ITU would thus effectively contribute to reducing the global digital divide.

Countries have different innovation systems, as the interaction of government institutions, firms, universities and research labs has evolved within different countries in various ways. Innovation dynamics is not only a matter of specialization and human capital formation. In the case of integrated countries (e.g., in the case of the EU, ASEAN or NAFTA), it is important to launch novel final products tailored to regional and global markets. What also matters at the level of the firm is the ability to adequately use the knowledge of specialized suppliers whose ability to develop novel subsystems is a crucial asset in the automotive industry of many countries. Moreover, using novel intermediate products imported from countries with successfully innovative firms is also an element of competitiveness in open economies. What matters more in the long run is the dynamics of the overall innovation system, which not only includes firms and their innovative suppliers but also specialized R&D firms, the innovative potential of researchers and labs at universities and the availability of modern infrastructure (roads, railways, airports, telecommunications). The incentives for innovation are partly intrinsic, but also partly in the form of expected rewards for which intellectual property rights and R&D promotion by government are important. Moreover, innovation is associated with risk so that sustained high innovation dynamics require favourable access to equity capital including venture capital.

Continental EU countries have traditionally relied much on the banking system which in turn financed most investments and innovation projects on the basis of collateral. In a modern digital service economy, the availability of collateral becomes a problem, however, since knowledge and software play an increasing role for existing and new innovative firms. Compared to the continental banking system, Anglo-Saxon capital markets – with a strong tradition in venture capital financing – are easier sources of financing innovative projects in the services sector. This could undermine the dynamics of modern industry in continental EU countries to the extent that innovative services are crucial inputs for manufacturing products or a key element for optimum after-sale service. To the extent that US multinational services companies invest in Europe or Asia, innovative services might become available despite weaknesses of the respective domestic services sector. However, high profits earned in innovative service firms will then accrue in the US which, in turn, could thereby strengthen digital US growth. In high wage countries of the EU, it seems to be quite important that nurturing innovative services not be neglected. It is an open question whether the expansion of the digital economy will particularly intensify high technology competition in OECD countries; if the digital world stimulates diffusion it undermines to some extent the incentive for innovation, since innovators can reap a Schumpeterian rent in world markets for a shorter period, at the same time there is an incentive for innovation leaders to accelerate the innovation race. If the expansion of the digital economy particularly stimulated high technology specialization, one may expect higher long term output growth since JUNGMITTAG (2004) has shown that high-technology specialization is a quasi-factor in production function. Thus innovation and skill upgrading could be quite important for OECD countries and NICs.

Theory of Innovation Policy

Why should government support R&D? This is obvious when it has a particular individual interest in innovations as is the case in the field of defence. Besides this special sector, it is only in the case of positive intersectoral or intrasectoral effects that R&D subsidies are adequate. Assume that the demand for an innovative product is given by the demand curve AZ (DD_0) – expressing private benefits – while social benefits are reflected by the demand schedule A'Z' (DD_{01}). As there are positive external effects of using the innovative product (or service), the optimum quantity will not be brought about by private markets themselves. The market would bring q_o while the optimum quantity is q_1; the latter would be produced if government reduced marginal costs by a fraction b so that there should be R&D subsidies. The price would fall to p_1 instead of p_0 in the simple market equilibrium. However, R&D subsidies would then amount to the area HGFD which can only be financed through (income) taxes, in turn shifting the AZ curve downward. Taxation would then impose deadweight losses – that is reduce economic welfare – unless the tax is on activities with a negative external effect. Since the latter case can be assumed to be relatively rare, the optimum R&D subsidy is slightly smaller than indicated by the subsidy rate b. Moreover, subsidization of R&D makes sense only if the increase in net welfare is higher than the costs of subsidization. These costs could ultimately include the costs of other sectors calling for equal treatment: read subsidization (while not showing positive external effects). In addition, there is a risk that government combines R&D subsidies with interference in the business sector which can cause efficiency losses. As a practical issue, one also has to look into the issue of granting subsidy payments or offering tax credits for R&D intensive firms. Subsidy payments appeal to the lobbying and rent-seeking efforts, particularly those of large firms. Tax credits are a superior instrument to the extent that R&D intensive small and medium-sized firms can also benefit from this relatively easily.

Government also has a role in the field of intellectual property rights. In the digital economy, intellectual property rights have come under pressure, because the violation of copyrights is rather easy (see, for example, the Napster trial).

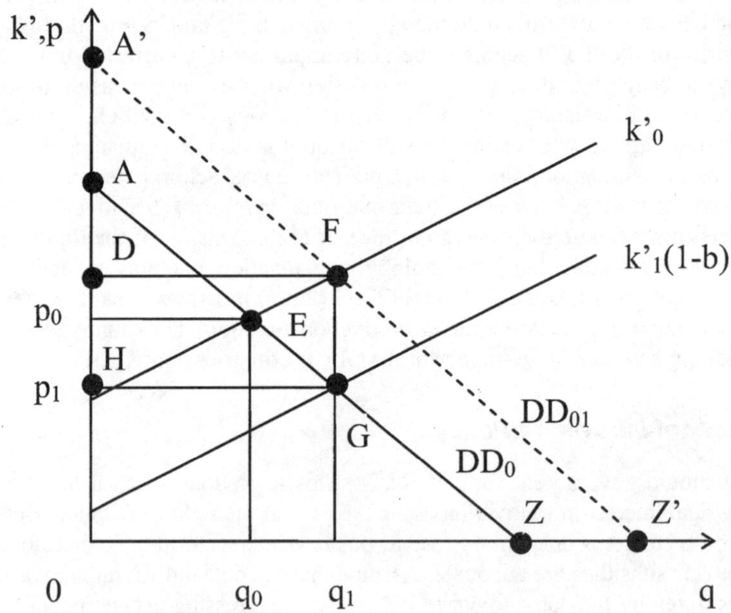

Fig. 10. Product innovation with positive external effects and R&D promotion

Network Effects

In the digital economy, there are more fields with network effects than in the traditional industrial economy. Network effects can be understood as an endogenous outward rotation of the demand curve (alternatively, as a rightward shift of the demand curve) in the process of network expansion. For example, the marginal utility of having access to the telephone network will increase for the initial users if more users – read potential communication partners – are switched on the network, at least as long there are no additional congestion costs. Similar considerations apply to advanced software and novel internet services. A monopolistic supplier facing the demand curve DD_o would impose the monopoly price p_2, which goes along with output q_0, under the standard Cournot monopoly solution. We assume for simplicity that a conservative monopoly firm would be neither willing nor able to exploit network effects. That is to anticipate that the dynamic demand curve – including network effects – is DD_2 and not DD_1. If government would offer a one-off R&D subsidy for process innovations (shifting the marginal costs curve from k'_0 to k'_1) to several firms under the condition that network effects be jointly exploited, the conditional competitive solution is point G. The positive welfare effect from the network effect is the area AEF plus the effect from the cost cutting, namely EFGH. An additional welfare benefit – not related to network effects – would come in the form of the area p_0EHp_1. Thus network effects represent a neglected and interesting field of R&D subsidies (for a broader partial equilibrium analysis also see appendix 1).

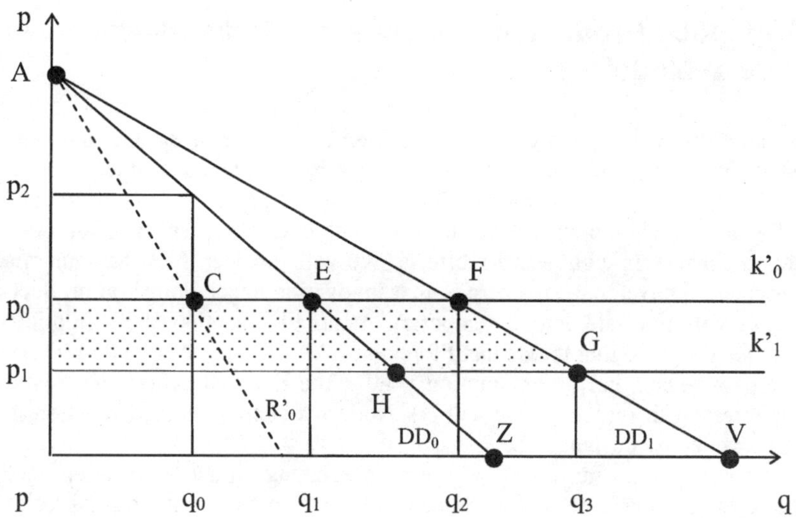

Fig. 11. Network effects and cost-saving progress

In a digital market economy, adequate support for R&D is quite important. However, since increased international networking is likely to create positive international innovation, spillovers there is an incentive for each government to underspend on R&D promotion (the latter should normally come in the form of tax credits or government subsidies for the wage bill of researchers – possibly with an absolute limit per firm as is the case in the Netherlands). Joint R&D promotion thus could become more important in a digital world economy. However, it will not be easy to organize efficient joint R&D programmes.

Restoring or maintaining equal opportunity in a digital social market economy certainly will require:

- serious efforts by governments to support digital learning in all schools
- special support and encouragement for the elderly
- explicit or implicit income transfers
- broadening the definition of universal services

To the extent that one is to shape rules for the global economy, the diverging models of the US free market economy and of the EU social market economy will face a critical stand-off. The global balance will tip in Asia, where the US and the EU are fighting for commercial and political influence. Since China is the largest market in Asia, it will be US-Sino and EU-Sino relations which could be decisive. The EU has a slight advantage only in mobile telecommunications; this might not be enough, however, to win the race for standards, rules and influence.

Appendix: Product Innovations and Network Effects in a Simple Model

Assume that willingness to pay depends on the degree of product innovation V, which in turn is proportionate to innovation efforts H, while production costs positively depend both on the quantity produced q and the innovation efforts H. In the following simple model, we will look at some interesting aspects of product innovations in sectors with network effects. We will consider the behaviour of an experienced innovator who is certain that innovation expenditures on product innovation will translate into a temporary monopoly position. For simplicity, we assume V=H and that the demand function is $q = H^{\alpha'} a/p$ (where a and α' are positive parameters; and by assumption α'>0. Note $E_{y,x}$ will denote the elasticity of variable y with respect to variable x). Note that such a demand function implies that revenue is constant since $R = pq = H^{\alpha'} a$.

Thus for the profit maximizing firm undertaking product innovations, we have the following expressions for the demand function function and for profit Q, respectively:

$$q = H^{\alpha'} a/p \tag{1}$$

$$Q = p(H)q - k(q,H) \quad (2)$$

Maximization of this function requires in a competitive framework:

$$dQ/dq = p - k'(q,H) = 0 \quad (3)$$

Hence

$$\underline{p = k'(q,H) \text{ under competition}} \quad (3\text{'a})$$

$$R'(q,H) = k'(q,H) \text{ under monopoly; with revenue } R = p(q.H)q; \quad (3\text{'b})$$

if revenue is constant – as in the case of an isoelastic demand function – we must simply minimize costs with respect to q while inserting from dQ/dH=0.

Furthermore

$$dQ/dH = \partial p/\partial H\, q - \partial k/\partial H = 0 \quad (4)$$

Inserting the demand function we have:

$$\partial p/\partial H\, H^{\alpha'} a/p = \partial k/\partial H \quad (5)$$

$$E_{p,H}\, H^{\alpha'-1} a = \partial k/\partial H \quad (6)$$

Hence the optimum product innovation effort H# is given by the necessary condition

$$H\# = \{[\partial k/\partial H]/aE_{p,H}\}^{1/1-\alpha'} \quad (7)$$

After choice of H#, the firm can choose output q in a way that the condition marginal revenue equals marginal costs is fulfilled. According to the above equation, the product innovation effort will be higher, the lower the marginal costs of innovation $\partial k/\partial H$, the lower the elasticity $E_{p,H}$ of the price with respect to innovation (efforts), and the lower the basic willingness to pay as expressed by a are.

How will network effects affect the elasticity of p with respect to H? If the market is characterized by network effects, we may assume that the elasticity of p with respect to product innovation effort H is lower than in a normal market (this effect could be estimated empirically), because network effects stimulate competition which in turn drives down the market price. Hence the optimum product innovation effort would be higher; every round of product innovation is a new starting point to escape the move towards low competitive prices. At the same time, there could be scale effects whose exploitation is facilitated in the presence of network effects. If there were only one firm, it would be easy to anticipate the positive network effect. However, the firm would not only charge a temporary monopoly price – which is a natural side-effect of the product innovation itself –, but it would charge a monopolistic price even in the long run. Moreover, it would adopt a lower rate of product innovation. With many firms in the market, the firm which launches a product innovation only has a temporary monopoly. As soon as there is broad imitation, the innovating firm can no longer fetch a monopoly price in the market. It must accept a price equal to long-term marginal costs (assumed to be constant in the subsequent graph where we show an initial monopoly price plus a situation with a product innovation in the context of monopoly pricing as an analytical starting point).

If there are quasi-scale economies in the sector considered, the marginal costs of product innovation will be negatively influenced by the size of the market, which in turn is positively affected by network effects. The latter thus brings about a positive external costs effect for all competitors. The assumption that network effects reduce the marginal costs of innovation can only be justified if the size of the market is a positive signal for attracting more researchers to R&D activities in which learning by doing or intra-sector spillovers plays a role. Note that under monopoly, we have price p^N for the product innovation both in the short term and in the long term. Under competition, however, it would be the lower price p_1 in the long run. Hence government should encourage competition and help prevent high barriers to entry from occurring.

Endogenous barriers resulting from strong network effects in industry (e.g., in the software sector) could also be a problem which would not always justify standard patent protection.

Our line of reasoning can also be seen form the following graph where we show an initial marginal costs curve k'_o and a new marginal cost curve k'_1 which is relevant in the context of the product innovation. The initial equilibrium point under monopoly and the initial demand schedule DD_o is point C and hence output is q_1 and the monopoly price p_M. Now we turn to product innovation. There will be a rise of the cost curve stemming from product innovation efforts, and this reduces consumer welfare. At the same time, there will be an upward rotation of the demand curve (see DD_1) reflecting the increased willingness to pay for the novel product; the latter effect is associated with higher consumer welfare. A positive network effect is reflected in the graph as a rightward shift of the demand curve (see DD_2). If there are quasi-scale effects, there will also be a downward shift of the k' curve. For simplicity, we can assume that this shift effect fully offsets the initial shift from k'_o to k'_1. Hence taking into account quasi-scale effects associated indirectly with network effects results in k'_o so that the long-term competitive equilibrium is point H, output q_5 and price p_o (without scale effects output would be q_2). Transitorily, there could be a monopoly situation with a monopoly price p_o and output q_6.

There are cases in which industry itself is able to exploit network effects fully (e.g., when leading firms in the sector agree on a new standard in the context of a product innovation). Setting standards in an environment of open interfaces of equipment allows for the exploitation of both network effects and scale effects. Modern electronics offers many examples of successful standard setting through industry itself. There are, however, well known examples of competing proprietary standards as well, as was the case with video recorders.

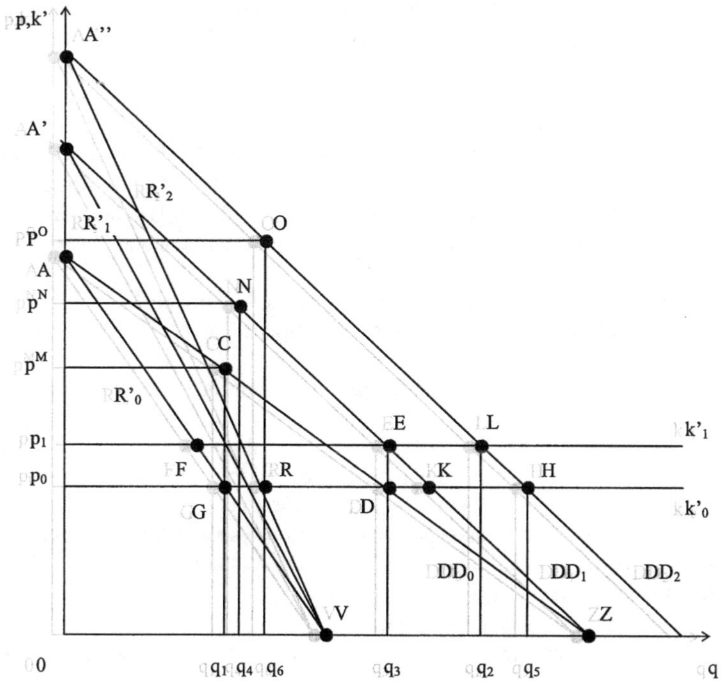

Fig. 12. Product innovation and network effects

References

AUDRETSCH, D., WELFENS, P.J.J., eds. (2002), The New Economy and Economic Growth in Europe and the US. Springer, Heidelberg and New York.

BAKOS, J.Y./BRYNJOLFSSON, E. (1999), Bundling Information Goods: Pricing, Profits and Efficiency, Management Science, December.

BAKOS, J.Y./BRYNJOLFSSON, E. (2000), Bundling and Competition on the Internet, Marketing Science, January.

BARFIELD, C.E., HEIDUK, G., P.J.J. WELFENS, eds. (2003). Internet, Economic Growth and Globalization – Perspectives on the New Economy in Europe, Japan and the USA, Heidelberg and New York: Springer.

BEESLEY, M. E., LITTLECHILD, S. C. (1989), The regulation of privatized monopolies in the United Kingdom. – Rand Journal of Economics 20, 454–472.

BORBÈLY, D. (2004) EU Export Specialization Patterns in Selected Accession Countries, EIIW Discussion Paper No. 116, University of Wuppertal, www.euroeiiw.de.

BRIGLAUER, W. (2006), Market Analyses Under the New European Communications Framework – Some Conceptual Issues, in WELFENS, P.J.J.; WESKE, M., eds., Digital Markets and Economic Policy Perspectives, Heidelberg and New York, forthcoming.

BUNTE, H.-J.; WELFENS, P.J.J. (2002), Wettbewerbsdynamik und Marktabgrenzung auf Telekommunikationsmärkten (Competition Dynamics and Market Demarcations in the Field of Telecommunications), Heidelberg: Springer.

CRANDALL, R.W. (2005), Competition and Chaos. U.S. Telecommunications since the 1996 Telecom Act, Washington DC: Brookings.

COMMISSION OF THE EUROPEAN COMMISSION (2004), European Electronic Communications Regulation and Markets 2004 (COM (2004) 759 final), Brussels.

DENIS, C., McMORROW, K., RÖGER, W., VEUGELERS, R. (2005).The Lisbon Strategy and the EU's Structural Productivity Problem. European Economy: Economic Papers No. 221, Brussels: European Commission.

EITO (2005), European Information Technology Observatory, Frankfurt/M.

EUROPEAN COMMISSION (2003), Staff Working Papers, European Innovation Scoreboard 2003, Brussels.

ITU (2003), World Telecommunication Development Report 2003: Access Indicators for the Information Society, Geneva.

JUNGMITTG, A. (2004), Innovations, Technological Specialization and Economic Growth in the EU, forthcoming in: International Economics and Economic Policy, Vol. 1, No. 2/3, Special Issue, Springer, Heidelberg – Tokyo – New York.

KNIEPS, G. (1998), Das neue Trassenpreissystem: Volkswirtschaftliche Vorteile eines zweistufigen Systems, *Internationales Verkehrswesen*, Vol. 50, 466-470.

OECD (2003), OECD Communications Outlook, Paris.

OECD (2005), Communications Outlook 2005, Paris.

OECD (2006), Telecommunications Regulatory Institutional Structures and Responsibilities, Paris.

OPTA (2005), Regulating Emerging Markets?, Economic Policy Note, No. 5, Amsterdam, mimeo.

STOCKDALE, D. (2003), The Regulation, Deregulation, and Nonregulation of Telecommunications and the Internet in the United States, in: BARFIELD, C.E.; HEIDUK, G. and WELFENS, P.J.J., eds., Internet, Economic Growth and Globalization, Heidelberg and New York: Springer, 249-274.

VAN ARK, B. (2001), The Renewal of the Old Economy: An International Comparative Perspective, STI Working Papers 2001/5, Paris: OECD.

VAN ARK, B., PIATKOWSKI, M. (2004), Productivity, innovation and ICT in Old and New Europe, *International Economics and Economic Policy*, Vol. 1, 2+3, Heidelberg: Springer.

VOGELSANG, I. (1998), Optimal price regulation for natural and legal monopolies. Paper prepared for CIDE seminar on the structural reform and regulation in the energy sector. http://www.cre.gob.mx/english/publications/researchindx.html.

WELFENS, P.J.J.; YARROW, G., Hg. (1996), Telecommunications and Energy in Systemic Transformation, International Dynamics, Deregulation and Adjustment in Network Industries, Springer: Heidelberg.

WELFENS; P.J.J.; GRAACK, C. (1996), Telekommunikationswirtschaft (Telecommunications Economics), Heidelberg: Springer.

WELFENS, P.J.J.; JUNGMITTAG, A. (2002), Internet, Telekomliberalisierung und Wirtschaftswachstum, Heidelberg: Springer.

WELFENS, P.J.J., PONDER, J. (2003), Digital EU Eastern Enlargement, EIIW paper No. 109, University of Wuppertal (www.euroeeiw.de), October 2003.

WELFENS, P.J.J. (2002), Interneteconomics.net, Heidelberg und New York: Springer, 2nd edition forthcoming.

WELFENS, P.J.J. (2004), Significant Market Power in Telecommunications: Theoretical and Practical Aspects, Paper presented at the ITU-conference Regulation of Telecommunications, Vilnius, October 5-7, 2004, forthcoming as ITU digital proceedings volume.

WELFENS, P.J.J. (2005), Information and Communications Technology, Regulations and Growth: Theory and Key Policy Issues, in: WELFENS, P.J.J., et al., Integration in Asia and Europe: Historical Dynamics, Political Issues and Economic Perspectives, Heidelberg/New York: Springer, 31-65.

WELFENS, P.J.J., ZOCHE, P., JUNGMITTAG, A., ET AL., (2005), Internetwirtschaft 2010/(Internet Economy 2010; study for the German Ministry of Economic Affairs and Labor), Heidelberg: Springer.

WELFENS, P.J.J. (2006), Innovations in Macroeconomics, Heidelberg/New York: Springer, forthcoming.

WELFENS, P.J.J. (2006), Competition and Regulation of Network Industries, The Economics of Rational Regulatory Policy, EIIW discussion paper No. 143, University of Wuppertal, January 2006, forthcoming.

Market Analyses under the New European Communications Framework - Some Conceptual Issues

Wolfgang Briglauer[1]

1 Introduction 63
2 Market Definition 65
3 TCT – What's that? 69
4 Competition Analysis 71
5 Regulatory Remedies 78
6 Concluding remarks 83
Appendix 84
References 88

1 Introduction

For electronic communications markets, a new regulatory framework was enacted and should have been transformed into national law by all member states of the EU by the summer of 2003. In fact, this new regulatory framework is composed of five different directives, stretching the goals of a harmonized regulatory approach within the National Regulatory Agencies (NRAs). On whole, the new regulatory framework is designed as to ensure the transition to general competition law. As the title suggests, our specific focus will be on the completely rearranged and much broader market analysis procedure within the new framework. In doing this, the paper is supplemented by the relevant legal documents. Accordingly, this paper makes various references to important passages in the Framework Directive, the Commission's SMP Guidelines[2] and IRG/ERG documents discussing methodological aspects.[3] Concretely, the directives (especially Articles 14-16 of the Framework Directive) in conjunction with the Commission's SMP Guidelines outline a three stage analysis process: preliminary definition of relevant communica-

---

[1] The views expressed are those of the author entirely and do not represent those of RTR or TKK. Professional thanks to Paul Pisjak and Christian Paiser; Very special thanks to Andy Hyacinth!!

[2] EUROPEAN COMMISSION, 2002. Henceforth referred to as the "SMP Guidelines".

[3] In practice, the independent regulatory authorities coordinate themselves on an informal basis in the Independent Regulators Group (IRG) where, among other things, common positions are worked out. For this purpose, the IRG maintains expert working groups to deal with different issues. Official IRG positions are voted on in the plenary meetings. Subsequently, IRG results are discussed together with the European Commission in the European Regulators Group (ERG).

tions markets, examination of effective competition on those markets and discussion of appropriate regulatory instruments based on any competition deficits. Although it appears to be practically useful to structure these stages sequentially, logical linkages, as it will be shown, conceptually exist among them.

Given this setting, our paper has two main aims.[4] Firstly, we want to clarify some conceptual problems embedded in each of the three analysis stages. Such a necessity exists even though some issues are authoritative in nature, and thus in a way exogenous for NRAs.[5] Secondly, we want to compile practical experiences made during the first round of market analyses, as these represented a paradigmatic change for all NRAs. Thus it seems worthwhile to reflect on ideas and experiences both on a conceptual and a practical basis.[6] According to this viewpoint, we always try to especially focus on feasibility.[7]

According to the overall analysis procedure, this paper is structured as follows: Chapter 2 addresses the stage of market definition and discusses main difficulties associated with the standardized "HM Test" for market definition. Chapter 3 explores the interface between the first and second stage of market analyses, the so-called Three Criteria Test as defined by the European Commission. Chapter 4 is devoted to the second stage, diagnosing the competitive situation (as described in Article 16 §1 of the Framework Directive), which is referred to here as "competition analysis." As mentioned in the discussion below, the measurement of effective competition implies the need for a decision regarding the existence of significant market power (SMP). As "competition analysis" brings along huge changes and efforts, a guiding reference model is created that should provide NRAs with a rational and holistic basis for decision-making and evaluation in the identification of SMP operators and effective competition on electronic communications markets. In doing so, the reference model is shown to be a specific regulatory modification of the traditional "Structure Conduct Performance" Paradigm. Conceptual aspects concerning definition approaches as well as general statements and interpretations regarding the application and evaluation of competition indicators endorse this chapter. Chapter 5 is based on the results of competition analyses as well as on identified competition problems. In this respect, the discussion of regu-

---

[4] For a longer German version of this paper cf. BRIGLAUER and PISJAK, 2005; this contains a supplementary description of institutional and procedural aspects.

[5] Regulatory struggling with diverse consultants' expertise provides an extra reason for this intention.

[6] Due to the author's engagement in official expertises on the fixed voice telephony markets, all the examples and explanations given unfortunately have a strong „fixed network bias". The major characteristic of communications markets, however, the local loop bottleneck may be best illustrated by means of fixed network markets; this bias should therefore be of minor importance.

[7] For instance, theoretical literature gives a lot of advice on optimal Ramsey pricing in communications markets. Yet when one comes to estimate elasticities of demand for (intrinsically horizontally and vertically related) communications markets, practical experience discloses an armada of severe empirical problems, rendering such projects merely impossible.

latory remedies works as an interface whereby, identified competition problems should be regulatory addressed efficiently, proportionately and in a least burdensome fashion. Thus questions arise as to the appropriate layer for tying up regulatory instruments, that is wholesale or retail and on guiding principles when it comes down to the design of adequate remedial measures. The final chapter concludes and summarizes issues and questions which are still open for further discussions.

2 Market Definition

A significant change seen in the new regulatory framework regards issues of market delineation. In this respect, the recommendation on the relevant product and service markets published by the European Union serves as a basis. Contrary to the old legal framework, the underlying methodology of market delineation is to be based on economic principles. The purpose of placing the central competence for market definition within the European Commission was to have, to the greatest possible extent, the same relevant product and service markets in all Member States and in this way to create a basic precondition for the standardisation of European regulation of communications. Moreover, the national regulatory authorities may define additional or other relevant product and service markets to take account of the specific circumstances on the communications markets of the respective Member State. Under the old legal framework, however, market delineation was normatively predefined by markets that were identified in the directives. Of course, such market delineation usually did not reflect economic circumstances.

According to the new regulatory framework, relevant markets should be defined in accordance with competition law methodology/principles. The legal basis for this is the Framework Directive (2002/21/EC) Art. 15 (3), the respective SMP Guidelines and the Commissions Recommendation on relevant markets.[8] With re-

[8] EUROPEAN COMMISSION, 2003; according to that recommendation the following markets are declared to be relevant (although disprovable by NRAs, see below), which we rearranged and clustered for sake of clarity:

Relevant fixed telephony markets at the retail level

1. Access to public telephone network at fixed locations for residential customers
2. Access to public telephone network at fixed locations for non-residential customers
3. Publicly available local and/or national telephone services provided at a fixed location for residential customers
4. Publicly available international telephone services provided at a fixed location for residential customers
5. Publicly available local and/or national telephone services provided at a fixed location for non-residential customers
6. Publicly available international telephone services provided at a fixed location for non-residential customers

Relevant fixed telephony markets at the wholesale level

7. Call origination on the public telephone network provided at a fixed location.

spect to methodological aspects, thc so called "Hypothetical Monopolist Test" has become state-of-the-art, and as such, part of the communications framework. The base case scenario describes a company which currently and in the future only offers one product/service within a defined area. If the company in question permanently increases the price (reasonable but significant), the focus is laid on the question of whether customers/suppliers can react adequately without having to accept huge efforts and costs. In other words, if it is possible and worthwhile for customers/suppliers to switch over to substitutes sufficiently, then the original product/service and the substitute(s) belong to the same market. In this case the autonomy of the company increasing prices for the original product is limited and such behaviour would not be profitable. This idea can be repeated iteratively for aggregation of the original product and its substitute(s) until a monopoly with autonomy for setting prices can be found. In the "real world", a reasonable but significant increase of price is approximated by 5-10 %.[9]

Although that idea sounds quite plausible, many objections have been put forward both by theorists and practitioners. As this has been done extensively, it is not our intention to replicate the main arguments made therein. Instead we want to focus on some less pronounced aspects concerning the HM Test itself as well as the overall market definition procedure.

To begin with, it must be emphasised that HM/SSNIP test conceptually encompasses both, demand and supply side forces. Nevertheless, American and European jurisprudence favoured demand side substitutability in the past (shifting supply side processes to the second analysis stage). However, the SMP Guidelines principally attach equal importance to both forces: "Assessing the demand side and supply side substitution provides a way of measuring the quantity of sales likely to be lost and consequently of determining the scope of the relevant market." (EUROPEAN COMMISSION, 2002, §41). Although this viewpoint is not

---

8. Call termination on individual public telephone networks provided at a fixed location.
9. Transit services in the fixed public telephone network

Relevant fixed data markets at the wholesale level

10. Wholesale unbundled access (ULL)
11. Wholesale bitstream access for the purpose of providing broadband services

Relevant leased line markets at the retail and wholesale level

12. The minimum set of leased lines (retail level)
13. Wholesale terminating segments of leased lines
14. Wholesale trunk segments of leased lines

Relevant (wholesale) mobile markets

15. Access and call origination on public mobile tel. networks
16. Voice call termination on individual mobile networks
17. The wholesale national market for international roaming on public mobile networks

Relevant broadcasting markets (irrelevant for our purposes)

[9] Due to the test procedure of the HM test, it is sometimes referred to as the SSNIP test as well, which is designed to explore the consequences of a (hypothetical) **S**mall but **S**ignificant **N**on transitory **I**ncrease in **P**rice.

persevered unambiguously throughout the SMP guidelines either,[10] a consequent application of the HM test rules out any conceptual separation for market delineation purposes. When the HM raises the relevant price, some of its customers will reduce consumption or will choose not to purchase at all and drop out of the market. Furthermore, demand side substitutability is determined by the extent to which customers of the relevant product under consideration regard other (similar) brands as an acceptable substitute. The closer these similarities are, the more consumers will switch to these other products. On the other hand, some firms already operating near to the candidate product[11] will alter their production facilities and supply similar products to other consumers (remaining in that market). From the consumers' perspective, it makes no difference if potential substitutes were pre-existent (prior to the initial price increase) or if they were supplied by firms operating near the market (in response to the initial price increase). Rather, consumers just decide on relative prices and other product characteristics; demand and supply side forces interact simultaneously.

Thus the full set of competitive forces will answer altogether whether the price increase will be profitable or not. However, the term "competitive" indicates that the stages of market delineation and market analysis are inextricably based on common grounds (and will always be intertwined logically to some degree). Nevertheless, conceptual key distinctions remain. For example, market definition is based (hypothetically) on a competitive price level as a point of departure, whereas competition analysis focuses on actual price patterns (of existing firms).[12] Beyond such reasoning, the sequential approach simply turned out to be practically useful.

Supply side substitution is not only a vital element in market definition, it also covers a wider application field than is commonly held for at least two reasons: First, there is a questionable separation with respect to the concept of "potential competition" put forth in the SMP Guidelines: "A third source of competitive constraint [next to demand and supply side substitution] on an operator's behaviour exists, namely potential competition. The difference between potential competition and supply substitution lies in the fact that supply side substitution responds promptly to a price increase whereas potential entrants may need more time before starting to supply the market. Supply substitution involves no additional significant costs whereas potential entry occurs at significant sunk costs." (EUROPEAN COMMISSION, 2002, §38). Thus, according to the SMP Guidelines the speed of reaction marks the significant difference. When referring to the contestability the-

[10] §51 for instance: "Proper delineation of the product market *may*, however, require further consideration of potential substitutability from supply side"; italics added by the author.

[11] For sufficiently flexible production capacities, it seems likely that potential supply side substitutes must be technologically near to the candidate market.

[12] Another form of analytical dependency is inherent in the residual demand elasticity itself: „In a static model, the elasticity of residual demand completely summarizes a firm's market power and determines its pricing.", CARLTON and PERLOFF, 2000, p. 610. Of course in a dynamic framework, things are growing more ambiguous, as the outcome is determined by the way one models the reactions (in prices or quantities) of rival firms.

ory,[13] however, this differentiation does not hold. Instead one finds that entry, as one of the underlying crucial assumptions, must take place quickly without the need of substantial and time consuming additional investment expenditures. Only this allows potential entrants to take advantage of any profit opportunities in the respective market. Without this the so-called "hit and run entry" mechanism would not work, neither potentially nor actually. If at all, the hypothetical nature of the mere threat of potential entry as opposed to actual supply side competition might be seen as a relevant distinction. Although the SSNIP test is designed to take into account actual changes in quantity (and profits respectively) on an empirical basis, due to severe informational limitations it frequently remains a qualitative thought experiment.[14] Moreover, appraising the actual/potential extent of supply side substitution seems to anyway be a rather hypothetic task in many situations. Thus these boundaries become blurred inevitably, which might be a reason for the SMP Guidelines reneging on the strict assignment concerning potential competition in footnote 24: "What matters however is that potential entry from other suppliers is taken into consideration at some stage of the relevant market analysis, that is either at the initial market definition stage or at the subsequent stage of the assessment of market power (SMP)." Although a large portion of (both academic and political) literature adopted this trichotomy of competitive forces, this is not the case throughout: "Two different situations can be distinguished. Supply side substitution from actual competitors and supply substitution from potential competitors." (SLEUWAEGEN, 1999, p. 11).

Second, the importance of supply side substitution is reinforced when one takes into account that demand side substitution alone would often lead to an impractical and meaningless definition of a very huge number of markets. For instance, in the case of fixed voice telephony call markets,[15] there is virtually no demand side substitution for each individual call and therefore each call would constitute a single market. In such cases, one often resorts to the "similarity of competitive conditions" argument,[16] according to which single services may be aggregated whenever competitive conditions are sufficiently homogenous. Not only has this approach seemed to be rather vague, it also brings together market definition and analysis in a hardly comprehensible way. Yet explicit reference to supply side substitution would be much more conceptually sound insofar as this would also be

[13] "Potential competition" and "contestable markets" are used interchangeably in economic terms.

[14] "Although the SSNIP test is but one example of methods used for defining the relevant market and notwithstanding its formal econometric nature, or its margins for errors, its importance lies primarily in its use as a conceptual tool for assessing evidence of competition between different products or services.", EUROPEAN COMMISSION, 2002, footnote 26. In a way there is an inevitable trade-off between inherent shortcomings of an informal approach and the practical and empirical problems of rigorous statistical methods. Probably a holistic approach, making use of all available information and methods (qualitative and quantitative) will address this dilemma most adequately.

[15] Markets number 3-6 in footnote 8.

[16] Cf. for instance, PLUM et al, 2000, pp. 32 or SMP Guidelines §56.

in line with the SSNIP test methodology. In fact, the "similarity of competitive conditions" essentially results in the analysis of supply side substitution processes. To come back to our example, there might be one common domestic market for fixed voice calls, since a hypothetical monopolist raising prices for a certain call from subscriber A to subscriber B might be easily disciplined (given nationwide ex ante specifications such as CPS/CbC; see below) by another operator who could enter this "market" (A→B) immediately by changing production capacities (from A→C to A→B) without any significant additional costs. Aggregation ends when technical, regulatory or economic conditions restrict supply side substitution.[17]

All in all, excluding or undervaluing supply side substitution in advance for market definition purposes is ill-founded. Rather, competitive forces stemming from the supply side are a vital element in the preliminary stage of market definition.

3 TCT – What's that?

The methods described so far are more or less standard tools accruing from competition law. Given the sector specific ex ante framework, however, the Commission introduced a further concept intended to allow for that principal difference. The Commission has developed a so-called "three criteria test" (TCT) that serves as a "filtering tool" when considering whether a market should be regulated by competition law alone, or whether the market is susceptible to ex ante regulation. Market delineation results should thus be subject to an additional requirement for ex ante regulation.

In its Recommendation, section 3.2, the Commission lays down three criteria (to be applied cumulatively) for identifying relevant markets:

- The presence of high and non transitory entry barriers
- Dynamic aspects: does the market tend towards effective competition (without ex ante regulation)
- Relative efficiency of competition law

The Commission has applied TCT when defining the markets in the Recommendation. This means that if NRAs do not deviate from the markets in the Commissions Recommendation, there is no need to discuss these criteria for the purpose of analysing whether a market is relevant for ex ante regulation. However, the first and the second criteria will inherently be taken into account in the SMP assessment process nonetheless. Therefore a related investigation will either be done when perfoming the market analysis, or it has to be done beforehand in those cases in which the definition of relevant markets and the market analysis are (for institutional or other reasons) separated exercises. Concerning the third criteria (in

[17] Cf. EUROPEAN COMMISSION, 2003, chapter 4.2.1 for the delineation of national and international markets.

conjunction with criteria two), it must be at least partly dealt with within market analyses. It is obvious that market analyses cannot be conducted independent of underlying ex ante regulations, as these crucially determine market outcome. The explanations in chapter 4 and 5 will show that this aspect in turn depends on the conceptual understanding of "self sustainability".

The Commission expects the NRAs to follow the same basic criteria and principles when identifying markets other than those appearing in the Recommendation and will also base future revisions of the recommendation on these criteria. Markets failing the test will not be part of subsequent recommendations from the Commission. Neither should such markets be identified as subject to ex ante sector specific regulation by the NRAs. Due to national/local circumstances, it is of course possible for the NRAs to find national markets complying with the three criteria, even though the corresponding market analysed by the Commission was not found to meet all three criteria.

The analysis carried out, may well show that there is effective competition in the market under consideration. If no SMP operator is designated, no remedies will be applied in that market. The Commission's pre-definition of a relevant market is therefore not a pre-definition of markets in which sector specific ex ante remedies must be imposed. As it seems, embedded quasi contradiction[18] indicates conceptual disagreements of TCT within the overall market analyses procedure.

The nature of the TCT scales back to a kind of replica competition analysis. Indeed in practice, it appears to be a rather reduced form and by no means a full scale analysis. Besides these virtual issues (making the distinction between TCT and full scale competition analyses), however, the practical problem arises if the stages of market definition and competition analyses are dealt with sequentially by NRAs (as in case of Austria) due to, for instance, general resource restrictions and/or (different) data requirements. In contrast, if both analyses stages are carried out simultaneously (despite organizational difficulties), the TCT becomes essentially redundant.

Accordingly, there is no theoretical justification for the TCT as an additional market analyses tool, since it is already part of in-depth competition analyses and remedy discussion. Significantly, there is also no correspondent within economic literature. Concerning the application of the TCT, the Recommendation provides only general and qualitative discussions, and by and large the analyses are not applied in a traceable form to the approaches suggested therein nor do the results seem to be based on actual empirical data.

Nevertheless, TCT enables harmonization strategies at a European level. It creates a built-in bias for NRAs to stick to the Recommendation list of relevant markets, as the TCT is only required in cases of deviation (with the Commission judging whether the empirical evidence shown by the NRA is sufficient). Yet in light of the logical interdependencies between market definition and competition analy-

[18] Such a contradiction need not necessarily exist due to the conceptual differences between "effective" and "self-sustaining" competition (cf. the "sustainability" discussion in the "Regulatory Options" category of Table 1).

sis (that exist naturally[19]) and seen pragmatically, TCT might also be interpreted in positive terms as a necessary point of departure and as such as a crude truncation of circularity (CTC).

4 Competition Analysis

It is a well-known fact that defining terms are always associated with a certain degree of randomness. This is especially true of attempts to put the extremely complex phenomenon of "competition" into words. Depending on the level at which competition is observed, its functional meaning and its interpretation in context, the relevant literature has yielded a great number of attempts to define this term. It is not our intention to mirror those approaches here. Instead, we want to sharpen our attention on the conceptual understandings embedded within the regulatory framework. Unfortunately in doing this, we find several different attributes accompanying the term "competition," which in sum do not put forth a consistent conceptual framework. It thus seems necessary to clarify conceptual understandings first.

The term used within the SMP Guidelines is "effective," whereby a connection is made with firms' individual market power. In the absence of effective competition at least one operator has market power to such an extent that "….from a structural perspective, and in the short to medium term, the operator has and will have, on the relevant market identified, sufficient market power to behave to an appreciable extent independently of competitors, customers, and ultimately consumers… ." (Cf. EUROPEAN COMMISSION, 2002, §30). Concerning this aspect, the framework overall makes a deterministic, clear cut and inseparable connection between effective competition (at the market level) and – in legal terms – SMP (or, economically, individual market power at the firm level): "…the notion of effective competition means that there is no undertaking with dominance on the relevant market. In other words, a finding that a relevant market is effectively competitive is, in effect, a determination that there is neither single nor joint dominance on that market. Conversely, a finding that a relevant market is not effectively competitive is a determination that there is single or joint dominance on that market." ("equalization principle", cf. EUROPEAN COMMISSION, 2002, §112).

Furthermore the label "effective" was used by regulatory institutions before the new regulatory framework came into effect. Representatively, we quote the closely related definition made by the Independent Regulators Group (IRG):

"Effective competition can be defined as the persistent absence of players with market power. In its essence market power is the ability to influence prices and of persistently enjoying higher profits than those enjoyed by firms which do not possess market power. In an effectively competitive market users are better off, because they are more likely to be provided a greater variety of products and/or ser-

[19] Cf. footnote 29 for an illustrative example.

vices of lower price and higher quality than they could be provided in a non competitive market." (IRG, 2001, p. 2.).

In the first part, competition is defined in terms of price theory and in negative terms as the absence of price setting power and supranormal profits. The second part emphasizes the role of the consumer or – roughly speaking – the dynamic merits of effective competition.

Thus far, things seem to be quite straightforward and explained in round terms. Yet things get unclear when one adds the different connotations depicted in the directives. In contrast to the SMP Guidelines, the Access Directive introduces the term "sustainable" several times:

Article 1, Access Directive: "The aim is to establish a regulatory framework, in accordance with internal market principles, for the relationships between suppliers of networks and services that will result in sustainable competition, interoperability of electronic communications services and consumer benefits."

Article 13 (2), Access Directive: "National regulatory authorities shall ensure that any cost recovery mechanism or pricing methodology that is mandated serves to promote efficiency and sustainable competition and maximise consumer benefits."

Although the general background of the framework as well as the underlying normative policy decisions suggest that sustainable competition has to be equated with infrastructure based competition (addressing the wholesale layer), such a conceptual conjunction cannot be found explicitly. Furthermore, sustainability can also be found in reference to end user markets. For example:

Recital 20, "Access Directive: In particular, operators with significant market power should avoid a price squeeze which results in the difference between their retail prices and the interconnection prices charged to competitors who provide similar retail services being inadequate in ensuring sustainable competition."

Article 12, Access Directive: "denial of access or unreasonable terms and conditions having a similar effect would hinder the emergence of a sustainable competitive market at the retail level, or would not be in the end user's interest."

Beyond this vagueness, sustainable competition has been given a slightly different connotation in regulatory discussions. In examining the effectiveness of competition in relevant communications markets, NRAs must answer the hypothetical question of whether self-sustainable competition without any form of ex ante regulation existed. For an example of this, we quote IRG: ".....a NRA should cease to promote competition when there is effective competition, provided market power is not likely to arise in the market under observation or in a related market as a result of removing regulation." (IRG, 2001, p. 4.).

Obviously, the question arises as to if effective competition implies that actually no form of ex ante regulation at both retail and wholesale levels is necessary or whether the ordinary ex post regime (as indicated by the TCT) is deemed sufficient to ensure the perpetuation of effective competition. Closer examination shows, however, that effective competition refers to the respective markets under consideration only. The following example should help to illustrate this point: If one defines effective competition as a form of competition that gets by without any form of ex ante regulation (on associated retail or wholesale markets), we will rarely find effective competition within the underlying fixed network retail markets.[20] Concerning voice calls, CPS/CbC[21] will (at least for the foreseeable future, depending on thc firmness of the local loop natural monopoly) be a vital and essential form of basic wholesale regulation. Yet given this form of ex ante regulation, retail markets could be assessed as being effectively competitive with respect to the observed market outcome (concerning the relevant competition indicators, see below). Effective, in this sense, should then be defined as independent of ex ante regulation within the same relevant market under consideration (most notably, ex ante price regulation). Otherwise (i.e., following the first approach), we could observe a situation one day in which the incumbent operator has only negligible market shares (and/or, lower prices) with no other operator being in a dominant position, but still the overall outcome could not be deemed effective, as there is still underlying CPS/CbC regulation. (By the way, the case seems quite analogous with respect to retail access and a potential – wholesale – resale obligation.[22]) Somewhat unclear is the Framework Directive (Recital 27) which states, "It is essential that ex ante regulatory obligations should only be imposed where there is not effective competition, i.e. in markets where there are one or more undertakings with significant market power, and where national and Community competition law remedies are not sufficient to address the problem." This might give rise however to contradictions in (plausible!) cases like those sketched above.[23] This is due to the particular inconsistencies which led to criticisms and to a basic denial of the equalization principle by some scholars.[24]

---

[20] This applies in like manner to all other markets that are affected by the natural monopoly characteristic of the local loop.

[21] Carrier Preselection (CPS) and Call-by-Call (CbC) are regulatory measures which ensure that customers can obtain call services from providers other than the (SMP designated) incumbent provider. Likewise, CPS and CbC obligations (required by the Universal Service Directive) ensure that alternative operators have access to interconnection services on a cost-oriented basis.

[22] The term "resale" ("wholesale line access" synonymously) refers to a package of services which is provided by the incumbent to its own end users in connection with the provision of POTS and ISDN lines. As a rule, these services are funded by the fixed charge to be paid by the end user. The fact that such services are available within the framework of a wholesale offer (also referred to as resale) shall enable the competitors that do not have their own access network infrastructure and have thus far provided only carrier services to offer their customers an overall package (tariff options) comprised of line access and carrier services in the field of fixed network voice telephony. In particular, the companies that currently provide voice telephony services mainly through carrier network operation (by means of Call-by.Call or Carrier Pre Selection) shall be able to create new innovative tariff models and establish their own attractive brands with improved customer service and marketing activities in the future, cf. RTR, 2003, p. 66. We will examine the regulatory impact of such a resale obligation more closely in chapter 5.

[23] Cf. also chapter 5 for further discussions on this issue.

[24] Cf. the criticisms made by the (German) MONOPOLY COMMISSION, 2001, 2003 and KOENIG et al, 2002.

The general uncertainty as to the meanings of competition terms is enhanced by the fact that other numerous attributes are also used in discussions of competition economics. For example, competition is also described as complete or incomplete, free, workable, functional, atomistic, innovation based, initiative based or potential. Some labels differ fundamentally in meaning and interpretation, while others describe the same or similar situations. Although the various terms are often associated with certain theoretical models, the terms "effective," "functional" and "workable" could be used synonymously, as these attributes reflect a similar basic concept. Due to (historic, see above) consistency, however, we will refer to "effective" competition even at the risk of overusing the term. In any case, it is purely a matter of standardizing the language used. On a positive and feasible empirical basis, we suggest a working definition based on the functions of competition for the overall context, borrowing from KANTZENBACH (1967) and CLARK (1940) and addressing the sustainability criterion as described above:

Effective competition is defined (in qualitative terms) as the simultaneous appearance of (sufficiently) optimal allocation, rapid and flexible adaptation to changing general circumstances as well as promotion of technical progress, efficiency based distribution of income, and the additional consideration of the role and sovereignty of consumers. Effective competition can only be assessed if it is solely (or at best) based on basic forms of wholesale regulation and on no forms of ex ante regulation imposed on the relevant market under consideration.

In terms of content, "effective" is used to indicate a reality-based deviation from the neo-classical ideal of perfect competition. According to this line of thought, perfect competition is a micro-economic construct which cannot be achieved, and the more important question is that of when the competition processes observed fulfill the functions required by our working definition. In addition, the price theory formulation disregards the dynamic component inherent to the concept of competition. The reference/organizational model for the measurement of effective competition in the relevant communications markets is described below.

“Qualitative” refers to the shortcomings of competition analyses based on quantitative methods (as discussed briefly below).

"(Sufficiently) optimal allocation" contains the neo-classical part of the working definition in that it refers to the welfare optima derived from the model of complete competition (exchange optimum, production optimum and overall market equilibrium – the optimum optimorum) which describe the state of economic efficiency. Considering this, our definition also incorporates the first part of IRG’s approach, whereby reference was made to market power in terms of the ability to raise prices above competitive levels.

"Adaptive flexibility" and the "progress aspect" reflect a dynamic understanding of competition.

The "emphasis on the consumer side" takes the corresponding focus of regulatory recommendations and legal regulations into account.[25]

Effective competition in call markets will typically depend crucially on basic forms of "wholesale" regulation and in these cases it can not be labeled infrastructure based.[26] As the discussion in chapter 5 indicates, basic forms of wholesale regulation might be aligned to retail markets for various reasons, but these will not change the nature of the remedies imposed. Finally, the sustainability concept encompasses the wholesale markets as well, but in a different meaning due to the vertical two stage market structure.

Being equipped with the conceptual understanding outlined above, one has to decide for an optimal framework for analyzing the effectiveness of competition. Although the SMP Guidelines contain specifications concerning methodological aspects, an overall analytical framework is missing. Literature taking up this task already exists, and it therefore suffices to summarize the main arguments and principles put forth therein.[27] Hereby the traditional "Market Structure – Market Conduct – Market Performance Paradigm" (MMMP)[28] forms the central framework for our reference model for the measurement of effective competition. The appendix contains an exemplary textbook-based representation of MMMP, adapted for the regulatory context that makes the significant difference. Although it can generally be applied to all markets, the "Regulatory options" category in the appendix emphasizes the ex ante nature of regulation specific to communications markets. In contrast, standard textbook representations of MMMP only show "Regulatory Options" as one of several explanatory factors under basic conditions like government policy, whereas "Regulatory options" are of considerable significance in competition analyses (and their results).[29] This is due to the interdependencies which naturally exist between the analysis stages in the assessment of ef-

---

[25] E.g., IRG, 2001, p. 4: "The importance attached to the concept of effective competition is better appreciated in terms of its outcome for users. Users are at the heart of the regulatory approach." Also cf. the emphasis on telecommunications consumers in Article 8 of the Framework Directive.

[26] The "regulatory options" category in table 1 of our appendix provides another presentation of the sustainability concept.

[27] Hence, this section draws heavily on BRIGLAUER, 2004 and BRIGLAUER, 2003, a longer German version of the former. A quite similar presentation can be found in KOENIG et al, 2002.

[28] MMMP was developed at Harvard by MASON, 1939 and later by BAIN, 1959.

[29] Naturally, regulatory issues also play an essential role in the preliminary stages of competition analyses. For example, wholesale unbundling (ULL, market number 10 in footnote 8) is regarded as a relevant market according to the recommendation list, but at the same time it exists solely due to the analogous remedy imposed on the incumbent SMP operator. Furthermore, the market definition and the remedy itself crucially shape the outcome of the competition analyses. Thus in the case of ULL, the sequential procedure of market analyses turns into a complete amalgamation. Although this is an exceptional case, it clearly points out that a logical and clear cut delineation of analyses stages cannot exist for any of the relevant markets. Nevertheless, such a sequential procedure proves to be practically useful und necessary.

fective competition and the discussion of regulatory options.[30] If competition analysis finds that effective competition is insufficient in the relevant markets, the question of appropriate and specifically suitable regulatory remedies (with regard to the actual competition deficits) arises simultaneously in connection with the resulting SMP position of one or more operators (as described in Article 16 of the Framework Directive and in light of the general regulatory objectives described in Article 8).

Moreover, this modified MMMP framework must not be understood as being in the early stages of its development. Its basic assumption was originally that of a monocausal interpretation beginning with an exogenous market structure which required certain market conduct, bringing about clear market performance. On the contrary, mutual dynamic interdependencies exist among MMMP dimensions and associated indicators. The fact that MMMP is to be understood as a complex system of correlations implies that the clinical delineation of sub-areas, or deterministic "pigeon holing" is not possible, nor would it correspond to the idea behind the model.

Therefore, the diverse objections to MMMP put forth in the past are not considered relevant in this context, as MMMP is not at all interpreted dogmatically but understood as an analytical framework or an organizational model: "While the structure conduct performance relationship is subject to debate, it nevertheless provides a useful framework for organizing and discussing a number of important concepts." (VISCUSI et al, 1995, p. 62).

At the same time, the interpretation of MMMP as a framework for the reference model is not exhausted in its organizational function; rather, it allows us to integrate a wide variety of competition theory concepts in the interest of a plurality of theories.[31] Thus the reference model is designed to be an orderly framework as well as a general and flexible analytical tool. Regardless of this general applicability, however, the competition situations in individual markets and the specific relevance and importance of various competition indicators must always be assessed on a case-by-case[32] basis.[33] Ultimately, the overall empirical material avail-

---

[30] For example, the regulatory decision to allow or enable indirect access operators (cf. footnote 21) has a sustainable and fundamental influence on the market structure; number portability could change demand side barriers substantially, "Resale" and ULL enable competition in access markets, etc.

[31] For an in-depth discussion, cf. BRIGLAUR, 2003, 2004.

[32] The individual factors will be assigned various levels of importance in different markets. A few (of many possible) examples can be used to clarify this point: The conduct strategy "price collusion" is assigned potential significance in oligopolistic (mobile) markets, whereas its significance is more hypothetical in strongly asymmetrical (fixed network) markets with a large number of participants. In retail markets, advertising and marketing strategies are typically essential parameters of entrepreneurial action, while in input markets these are usually (or sometimes) seen as negligible activities. In connection with this, we can assume that the transparency of information in input markets will generally be substantially higher than in downstream retail markets. Potential competition will be considered far more significant in long distance services in fixed linked voice telephony services (seen gradually and) in relation to the "markets" for mobile termination.

able must be interpreted and assessed on the basis of experiential knowledge as well as economic theory. Decision makers will not be able to avoid assigning priority to certain competition indicators (at least in terms of significance) in light of individual market conditions. "It is quite possible that the criteria used will not all indicate a straightforward conclusion: a final judgment will involve weighting the various criteria and inevitably greater weight will be placed on some rather than others, with some criteria having relatively little importance for some markets." (IRG, 2001, p. 11). Naturally, such a form of weighting necessarily involves value judgments in regulatory/competition policy and an element of decision making which cannot precisely be verified. An explicit and generally-applicable weighting scheme, however, is in no way a viable alternative: "Even a few competitive activities make it impossible to set forth a measure of competitive intensity to which all will agree. Even if one could measure competitive intensity along each and every single dimension of competition, our inability to convert units of competitive intensity from one dimension to another makes the general intensity of competition ambiguous and even meaningless." (DEMSETZ, 1995, p. 142). This also corresponds indirectly to IRG's position: "Clearly it will not be appropriate to use the indicators mechanistically and in making its assessment a regulator will need to consider the relevance of each indicator in regard to the circumstances of the market segment being assessed." (IRG, 2001, p. 11).

From a legal standpoint, insufficiently effective competition is (implicitly) equated with the existence of significant market power on the part of at least one operator ("equalization principle"). The complex analysis tool used to identify effective competition on relevant communications markets has already been discussed. The assessment of significant market power, on the other hand, is carried out (by definition) at enterprise level. According to the SMP Guidelines, criteria taken from the decision making practice of European courts and the European Commission are to be taken into particular consideration when evaluating SMP positions (§ 78 for "single dominance", § 97 for "collective dominance").

The discussion above and a comparison of table 1 in the appendix reveal that all of these criteria (in § 78 and § 97) are already taken into account in our more comprehensive and systematic reference model. A large part of the criteria mentioned can be assigned to or dealt with in the market structure chapters. In any case, the model framework used for competition analysis – taken as a whole – also creates the foundation for the identification of SMP and results in the decision as to whether a SMP position exists. In its results, a competition analysis should provide information on both the intensity of effective competition and the related individual market power/SMP positions.

Under no circumstances are these criteria lists to be regarded as a complete checklist to be worked through step-by-step. Rather, the individual factors are to be understood as indicative. As there were and still are opinions which argue that market analyses should be conducted mechanically and legally by means of a pre-

[33] However, this does not rule out the validity of the inclusion of pattern forecasts nor the holistic view of relationships across markets.

defined and static list of indicators,[34] we will finally demonstrate that this is not at all in line with the intentions of the new regulatory framework. The SMP Guidelines explicitly (and implicitly) express that the indicators listed cannot be seen as closing (§ 78): "In that regard, the following criteria can also be used to measure market power...... These criteria include amongst others..." With respect to the joint dominance criteria listed, § 98 states: "Annex II of the Framework Directive expressly states that the above is not an exhaustive list, nor are the criteria cumulative. Rather the list is intended to illustrate the sort of evidence that could be used... ." Further assertions can be found in the relevant legal documents. Seen economically, it is even more apparent that the listed criteria can only be understood as indicative insofar as any distinction of relevant, potentially relevant or non-relevant criteria seems at most questionable. For instance, there is no economic explanation for why "Excessive Pricing", "Pricing" or "Price trends" is not included in §78 as a relevant indicator. Indeed, in predominantly homogenous communications markets prices may be seen as one of the most important key indicators, whereby it is economically reasonable to make use of any information conveyed in price patterns with respect to their dynamic pattern, their relation and their differentials to other operators or to costs. Furthermore, what about the difficulties arising from interdependencies and redundancies within indicators: "Control of infrastructure not easily duplicated" and "vertical integration" of the incumbent firm within the industry in conjunction with the "overall size of the undertaking" are mirrored within considerable economies of scope and scale. Considering demand and supply side barriers (although not explicitly mentioned in the single dominance list), it is evident that most of the criteria refer to them.

An economic discussion clearly indicates that a spontaneous, sporadic and isolated view of the indicators identified as relevant may yield a highly-distorted picture of the actual competitive situation. In this tenor, the SMP Guidelines state in § 78: "Therefore, NRAs should undertake a thorough and overall analysis of the economic characteristics of the relevant market before coming to a conclusion as to the existence of significant market power." Only in an orderly framework in a systematic overview of all (in given markets potentially relevant) competition indicators is it possible to describe and assess competition processes suitably. This requires a flexible application and interpretation but not a predefined list of fixed indicators which implicitly defines all other indicators not mentioned in the list as irrelevant. The reference model described above is intended to provide NRAs with a rational and holistic basis for decision-making and evaluation for the identification of effective competition on electronic communications markets.

5 Regulatory Remedies

It is widely accepted that the successful transformation of monopolistic electronic communications markets into competitive ones requires regulatory intervention

[34] Cf. footnote 5.

that goes beyond general competition law. Without ex ante regulation, sustainable/effective competition is not likely to emerge, or at least not at a rapid pace. Ex ante regulation was established within the telecoms sector in an attempt to reflect the asymmetry between established undertakings (incumbents) and new entrants. Asymmetric regulation needs to be reduced as competition evolves. The purpose of asymmetric ex ante regulation is generally fulfilled when infrastructure based competition exists in the relevant markets.

Beyond this general and widely accepted background, problems arise when it comes to implementation of the third and final layer of market analyses.

Competition deficits identified in market analyses and the discussion of optimal regulatory remedies based on even those problems define the interface between stage two and three of the market analyses procedure. To achieve this, competition analyses have to be carried out considering the existing ex ante framework (in relation to the respective markets under consideration), whereas stage three should elaborate on an optimal ex ante framework. Though conceptually evident, experience shows that it is worth emphasising this point.

Assistance on a generic layer is provided by the ERG common position document on the appropriate remedies in the new regulatory framework (ERG, 2003). This document first develops a generalization of competition problems, describes the remedies and guiding principles available for NRAs, and then applies those to reveal how to pair certain competition problems with regulatory ex ante remedies. Of course in practical application, actual problems arise which in turn depend on the specifics of the individual member state and are thus left necessarily to the NRAs. Therefore we want to sharpen our focus on some regulatory aspects which should be of general interest.

Although depending on national law implementation, probably all NRAs have been confronted with difficulties concerning the appropriate market assignment of remedies. Once again, legal and economic requirements might be in conflict (e.g., in the case of CPS/CbC). Due to the nature of the CPS/CbC obligation, one has to refer to it as a basic form of (essential, cf. the discussion above) wholesale regulation. But the universal directive (article 19) associates CPS/CbC obligation with an "..undertaking notified as having significant market power for the provision of connection to and use of the public telephone network at a fixed location...". That is, the framework not only links CPS/CbC to the retail layer but unfortunately to all (i.e., access and usage) retail markets. Economically, CPS/CbC obligation should depend on a SMP assignment on the call origination (wholesale) market, since CPS/CbC refers to standard products of the call origination market. Due to legal linkage to the retail layer, however, this is not possible. Looking at those markets, it is obvious that the market power in the call origination market is de facto mirrored in the retail access markets since both express the (still existing) natural monopoly of the "local loop".[35] Thus, the obligation has to be assigned

[35] The local access networks still have a sub additive cost structure, i.e. one provider of infrastructure can handle the entire demand in the local loop area more cost efficiently than two or more. As long as alternative access technologies such as energy supply networks and Wireless Local Loop (WLL) still lack technological sophistication to reach the state

with SMP positions on access markets (alone), while of course differentiating between residential and non-residential markets. In no case would it be meaningful to interpret article 19 literally and impose CPS/CbC only if an undertaking is found to have SMP on all access and call markets. In accordance with our definition, call markets (might) pass the sustainability test (with regards to ex ante obligations underlying even those markets) and might therefore be called effectively competitive. At the same time, it is evident that infrastructure based competition is still not foreseeable as opposed to retail mobile markets.

Overall, the solution proposed might be seen as a compromise accounting for both legal and economic demands ("good second best"). At this point, we wish to offer another important example reflecting assignment difficulties in a slightly different light. Already initiated under the old regulatory framework, resale of subscriber line (or wholesale line rental) has received ever-growing attention among NRAs and was also strengthened by the new framework which now explicitly accounts for resale as a remedy. In regards to obligations of access to (therefore addressing the wholesale layer) and use of specific network facilities, article 12 of the Access directive states that "Operators may be required inter alia (d) to provide specified services on a wholesale basis for resale by third parties;" According to the new regulatory framework, the Access Directive generally refers to (ex ante) wholesale obligations. Furthermore, the resale of subscriber lines clearly refers to competition deficits on the (monopolistic) access side. Yet consulting the recommendation list, there is no appropriate wholesale access market. In a way second best, SMP assignments on the retail access markets are deemed to be the point of origin. Again, an adverse situation might arise (showing the second best character) in which the findings indicate effective competition precluded by definition, as those markets would then be dependent on ex ante regulation (and hence unable to pass the sustainability test). There have been a lot of discussions among NRAs on exactly that issue with no final conclusion generally accepted. Conceptually, it would seem to be most consistent to define a relevant wholesale access market and impose the resale obligation on that market.[36] At least this would be regulatory consistent with respect to the other wholesale access markets,[37] since practically the same paradox situation we already mentioned concerning the CPS/CbC obligation might otherwise arise. Due to the successful implementation of resale, market outcomes might one day indicate effective (service-based) competition on (both) retail access markets, but deregulation would be forestalled

of marketability and while cable TV networks have substitution potential only in conurbation areas, the natural monopoly in the local network will largely prevail. Of course, the mobile communications sector as a whole has to be regarded as substituting the fixed network to a certain extent. At the moment, however, no sufficiently large substitution effects can be observed, even though this issue will be further seen and more closely examined in future market analyses. Cf. RTR, 2003, p. 159.

[36] With respect to residential and non-residential markets on the retail layer, NRAs will have to consider whether a corresponding differentiation would be meaningful at the wholesale access layer.

[37] Market number 10 and 11 in footnote 8.

since the overall result depends on ex ante regulation in exactly these markets. This form of contradiction indicates that resale is in essence a basic form of wholesale regulation.

Apart from this, we wish to refer to some of the general principles underlying the new framework which should guide NRAs in designing optimal ex ante regulation. Based on the aims of the directive (most notably framework directive article 8 as well as access directive article 8), these principles are consolidated in ERG's common position.

A core principle embedded in the new framework addresses the desirability of incentives for both infrastructure investment and infrastructure competition. Fortunately, ERG supplements this principle with the crucial aspect of replicability.[38] Infrastructure is termed replicable if rolling out of competing (non incumbent) infrastructure is technically and economically feasible over the projectable future. (Only) In case of replicability, NRAs should favour ex ante remedies in line with infrastructure based competition. At the bottom of this principle is the vision of a continuous transition path from monopoly towards ordinary ex post competition, with ex ante regulation being a necessary intermediate phase. The so-called "ladder of investment" emblematises this idea. Yet again, when confronted with real market outcomes we do not observe such a continuous development. In fact as Figure 1 indicates, there have been only minor changes concerning the initial natural access monopoly since liberalisation started (in 1997/1998). According to international experience, infrastructure based competition (labelled as "Direct access" in Figure 1) is around 6%. The essential point is that there has been no significant increase in the past and that there is still a huge gap to pure infrastructure based competition.[39] Figure 1 does not distinguish between direct access based on cable infrastructure (provided by alternative operators) and unbundling activities. The latter are widely identified as infrastructure competition, since they require much more investment expenditure compared with pure reselling activities.[40] In general, however, the unbundling market itself is nothing but an ex ante obligation. Given the incumbent's dominating market position regarding the local

---

[38] Even though the economic importance of investments in telecommunications infrastructure is undisputed, the conclusion – the more investments, the better – would be wrong, because from an economic point of view, the extent of investments in infrastructure can be either too high or too low. Increasing investment expenditures in the telecommunications industry cannot be regarded as positive at every given point of liberalisation and in every geographical dimension. The very fact that there are huge overcapacities concerning national and international backbone networks indicates the danger of over investments as compared to under investments.

[39] Obviously this result is in contrast to the mobile sector, in which several fully vertically integrated operators typically exist.

[40] To effect unbundling, the respective subscriber's local loop is (electrically) connected to the network of the unbundling partner in a separate room (colocation room). For this purpose, the unbundling partner has to make considerable investments in the adaptation of the collocation room, in backhaul as well as in separate switching equipment. RTR, 2003, p. 186.

loop, it is hard to imagine that unbundling activities would persist to the same degree under the ex post regime (with no incumbent obliged to provide wholesale unbundling, with no regulator defining monthly rates and other general terms). Given our understanding of infrastructure based competition, whereby competition gets by without any form of ex ante regulation, it would thus be more reasonable to describe ULL as a high end form of service competition. Retail competition based on ULL might thus be effective but not infrastructure based.

Furthermore, both direct access components have not yet been provided nationwide, rather only in user specific (residential vs. non-residential) and regional (urban vs. rural) areas.

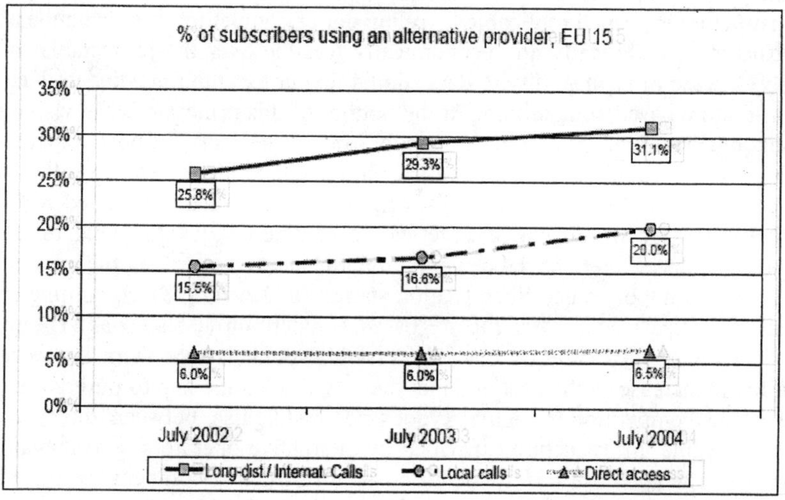

Fig. 1. The "Dynamics" of Local Loop Infrastructure Competiton – "per aspera ad astra"[41]

According to the "ladder of investment," access prices should be increased in the lapse of time to induce further investment activities. This involves some serious risks. First, it generally implies distortions (endangering the consistency of existing access price structure) given cost orientation and a static viewpoint. Second, it also involves high uncertainty concerning future market activities or technical developments. A third risk is that distortions favouring infrastructure competition might be more detrimental to social welfare than regulatory policies favouring

[41] Cf. the market overview in Annex 3, chapter 3, figure 21 of the tenth Implementation Report (issued by the EUROPEAN COMMISSION, 2004). In addition, the UK situation constitutes a major outlier (19% direct access via alternative providers) distorting the overall picture. Furthermore that British local loop success is not primarily due to regulatory intervention but rather due to historic developments; cf. KOENIG et al, 2002, p. 186.

service competition.[42] The ratio between infrastructure and service competition, implicitly depicted in figure 1, underlines our third point on empirical grounds. Considering these aspects, ERG suggests the following: "Thus, if the NRA is uncertain as to whether replication is feasible, it should maintain a neutral stance and continue to monitor the market, to firm up its view as to the likelihood of replication." (ERG, 2003, p. 68). Due to its importance, this should be a principle of equal rank. Furthermore, ERG accentuates the complementary role of service competition as a precondition for long term infrastructure competition (ERG, 2003, p. 68). Thus, in case of doubt there are good reasons for retaining a cost orientated price structure corresponding to a static perspective. Dynamic access pricing is even more informationally demanding which involves high uncertainties, and should therefore be restricted to cases in which replicability seems likely in the foreseeable future.

6 Concluding remarks

One of the most distinctive features of market analyses emerging from the relevant articles of the new regulatory framework (as compared to the old one) refers to the much wider economic focus in conducting each market analysis stage. Obviously, this requires a lot of additional time and effort. The overall project will therefore only pay off if the unique feature, the holistic economic perspective, receives due attention.

The very fact that most NRAs cluster relevant markets – although in different ways due to resources, experiences or other institutional reasons – may be seen as evidence for inherent linkages among relevant markets and the need to adopt a holistic perspective.

Although NRAs in principle have discretionary power in drafting SMP decisions with respect to market definition and the assessment of SMP, the European Commission may as a last resort prevent a decision by the regulatory authorities by exercising a veto based on Article 7 of the Framework Directive. In case of such a veto decision, all NRAs are required to comment on the decision at hand. A low degree of effective participation indicates some institutional inefficiency; accordingly, a revision process is currently underway within IRG.

Whereas the assumption of sufficiently rational NRAs maximizing overall welfare within European Member states seems to be plausible,[43] the European Commission seems to maximize overall harmonization under consultation mechanism

---

[42] For a demonstrative example referring to the relationship between wholesale line rental and unbundling activities, cf. BEARD et al, 1998, pp. 315-326.

[43] All market analyses conducted under the new regulatory framework are subject to the (implicit) assumption of sufficiently rational consumer behaviour. All the more, our assumption appears to be reasonable since NRA's represent a dedicated capacity for permanent market monitoring.

in Article 7. It is open for discussion as to how far these aims are at odds or might be in line with each other.

Appendix

The appendix depicts a representation of the modified MMMP (adapted for the regulatory context). Furthermore, Table 1 tries to combine the functional outcome associated with effective competition (as defined in chapter 4) with the organizational framework of MMMP (as far as this was possible). Apparent difficulties in drawing a clear distinctive line reemphasise inherent linkages. To constitute a claim of completeness would be self-contradictory, but the main competition indicators and forces at work should be included.[44]

[44] Table 1 is a compilation made of the presentations in KOENIG et al, 2002 and BRIGLAUER, 2003, 2004.

Table 1. Reference model diagram

General conditions	Market stage discussion (S curve, etc.) Overall economic situation, business cycle aspects (where applicable) Fundamental provisions in competition law and regulatory policy; general institutional and regulatory conditions Previous regulatory decisions Access to capital markets / financial resources, international ties (Industry) standards ...
Market structure	Number and distribution of providers (or demanders in wholesale markets, countervailing buyer power) Development of market shares, concentration measures (HHI, based on revenues, minutes, subscribers, etc.; consolidation processes) Market entry and exit barriers (supply side barriers) (licenses, technology, product differentiation, etc.) Horizontal and vertical interrelationships among input and retail markets (and associated structural competition problems, e.g. leveraging) Description of international ownership (if necessary) ...
	Optimal allocation Concentration according to economies of scale; vertical and horizontal integration according to transaction costs and economies of scope; amount of product differentiation according to consumer demands and costs of differentiation; free and easy entry Adaptive flexibility / innovational progress No general prediction since the empirical evidence on Schumpeterian theory is ambiguous

Market conduct	Pricing policy (current rate situation, development of rates in reference period, relation to relevant costs, strategic pricing, reference to interconnection fees (where applicable), bundling strategies, pricing strategies influenced by regulation) (Price) collusion Quality based competition Investment activities Advertising and marketing – product differentiation[45] Distribution policy R&D – innovation activities …
	Optimal allocation Prices equals (marginal)[46] cost; Informational advertising; no abusive behavioral strategies; no excessive/collusive pricing Adaptive flexibility / innovational progress Investment activities and capacities according to demand development and durability of assets; successful market implementation of new products and technologies

[45] Generally, Table 1 addresses both wholesale and retail layer; however, some aspects are primarily considered relevant in retail markets, advertising and marketing.

[46] For inherent estimation problems and relevant cost concepts, cf. BRIGLAUER and SCHWARZ, 2006.

Market performance	(Primarily) From the consumers' perspective: International benchmarking Demand side barriers Innovation, quality and product variety Infrastructure and quality Information for customers / market transparency … (Primarily) From the enterprises' perspective: Performance discussion (rate of return, Price Cost Margins – static (allocation and technical) efficiency, etc.) …
	Optimal allocation Efficient production of goods, no X-inefficiency or distortions; no economic market power on demand or supply side and supranormal profits respectively Adaptive flexibility / innovational progress Technical progress and innovation according to costs and consumer preferences; steady growth of demand, no fluctuations due to slow going adjustments after changes in demand or supply parameters; free and easy market entry Consumer sovereignty Freedom of choice among a variety of products (priced at cost of production); variety and quality according to consumer preferences

	previous decisions and reason statements
	Fundamental provisions in competition law and regulatory policy; general institutional and regulatory conditions
	...
Regulatory options	Sustainability Tests
	Underlying and/or proposed regulatory instruments (on even that market) remain essential – no effective competition
	Removal of underlying remedies (on even that market) would not alter the sufficiently competitive environment – competition is effective (and sustainable with regards to ex ante regulation within the respective relevant market)
	Removal of any ex ante regulation would not change effective competition in respective market – market outcome is effective (and sustainable with regards to any ex ante regulation; thus competition is also infrastructure based; ex post regime is sufficient)

References

BAIN, J. (1959), Industrial Organization, New York: John Wiley and Sons.

BEARD, R., KASERMAN, D., MAYO, J. (1998), The role of resale entry in promoting local exchange competition, Telecommunications Policy, Vol. 22, 315-326.

BRIGLAUER, W. (2003), Generisches Referenzmodell für die Analyse relevanter Kommunikationsmärkte – Wettbewerbsökonomische Grundfragen, Bad Honnef: WIK Diskussionsbeitrag, Nr. 243.

BRIGLAUER, W. (2004), Generic reference model for the analysis of relevant communications markets: fundamental competition issues, info, Vol. 6, No. 2, 93-104.

BRIGLAUER, W., PISJAK, P. (2005), Marktanalyseverfahren im neuen Rechtsrahmen – Konzeptuelle Überlegungen und erste Erfahrungen, in Lichtenberger, E., Ruhle, E.-O., eds., Die Regulierung der österreichischen Telekommunikationsmärkte im neuen Rechtsrahmen, Reihe Regulierungsrecht und Regulierungsökonomie, Band 5, Köln: EUL-Verlag.

BRIGLAUER, W., SCHWARZ, A. (2006), Can the HM Test be brought to an end?, netnomics (forthcoming).

CARLTON, D.W., PERLOFF, J.M. (2000), Modern Industrial Organization, Addison-Wesley.

CLARK, J.M. (1940), Toward a Concept of Workable Competition, American Economic Review, Vol. 30, 241-256.

DEMSETZ, H. (1995), The Economics of the Business Firm seven critical comments, Cambridge University Press.

EUROPEAN COMMISSION (2002), Guidelines on market analysis and the assessment of significant market power, 2002/C 165/03, "SMP Guidelines".

EUROPEAN COMMISSION (2003), Recommendation of 11 February 2003 on relevant product and service markets within the electronic communications sector susceptible to ex ante regulation in accordance with Directive 2002/21/EC of the European Parliament and of the Council on a common regulatory framework for electronic communications networks and services, OJ 8.5.2003 L 114/45.

EUROPEAN COMMISSION (2004), European Electronic Communications Regulation and Markets 2004, "10th Implementation Report", available at the website of the European Commission.

ERG, European Regulators Group (2003), Common interpretation of the SMP concept for the new regulatory framework, available at the website of the European Commission.

IRG, Independent Regulators Group (2001), "Principles of implementation and best practice on effective competition in electronic communications markets".

KANTZENBACH, E. (1967), Die Funktionsfähigkeit des Wettbewerbs, Göttingen.

KOENIG, C., VOGELSANG, I., KÜHLING, J., LOETZ, S., NEUMANN, A. (2002), Funktionsfähiger Wettbewerb auf den Telekommunikationsmärkten: Ökonomische und juristische Perspektiven zum Umfang der Regulierung, Schriftenreihe Kommunikation & Recht Band 16, Heidelberg: Verlag Recht und Wirtschaft.

MASON, E.S. (1939), Price and Production Policies of large scale Enterprise, American Economic Review, Vol. 29, 61-74.

MONOPOLY COMMISSION (2001), Wettbewerbsentwicklung bei Telekommunikation und Post 2001: Unsicherheit und Stillstand, Sondergutachten, Bonn.

MONOPOLY COMMISSION (2003), Telekommunikation und Post 2003: Wettbewerbsintensivierung in der Telekommunikation – Zementierung des Postmonopols, Sondergutachten, Bonn.

PLUM, M., SCHWARZ-SCHILLING, C. (2000), Marktabgrenzung im Telekommunikations- und Postsektor, Bad Honnef: WIK Diskussionsbeitrag Nr. 200.

RTR, Austrian Regulatory Authority for Broadcasting and Telecommunications (2003), Communications Report 2003, available at www.rtr.at.

SLEUWAEGEN, L. (1999), Competitive distortions and state aid to firms. How to define the relevant market, Working Paper, available at europa.eu.int/comm/enterprise/ library/ lib-competition/doc/market_definition.pdf.

VISCUSI, W.K., VERNON, J.M., HARRINGTON, J.E. (1995), Economics of Regulation and Antitrust, Cambridge: MIT Press.

Liberalisation of Telecommunications and Innovation Dynamics in Denmark

Morten Falch and Anders Henten

1 Introduction 91
2 The Danish Telecom Market 93
3 Innovation in the Danish Telecom Sector 97
4 The Role of Domestic Demand 100
5 Denmark as a Test Market for Telecom Products 102
6 The Role of Competition 106
6.1 The Implications of Competition on Innovation Activities 106
6.2 The Implications of Innovation on Competition 110
7 Empirical Evidence 113
8 Conclusions 115
References 116

1 Introduction

Definitions and Delimitations

The paper examines how the liberalisation of the Danish telecom market has affected innovation activities. The paper defines the term telecom market as the market for telecom services. The market for telecom equipment is not included here, although interfaces between the operators and the equipment manufacturers are part of the examination.

In a study of the competition between individual operators on a national market and the role of innovations in this context, it is most relevant to reserve the innovation notion to the first operator(s) introducing the new technologies or methods. It does not have to be new in an international context. The innovative operator is the one which is the first to introduce a new product or process on the national market (e.g., the first to implement ADSL) even if ADSL technology has not been developed in the country in question. In the present paper, there is focus on both issues, that is the general innovativeness on the Danish market and on the role of innovations in relation to competition on the Danish market.

The innovation concept, in principle, includes product, process, billing and marketing innovations and the like. However, a large part of the innovations, which in the daily activities actually affect the market shares of operators, is related to billing methods and smaller product and service improvements. A known example of such innovations is the prepaid card on the mobile market in the late 1990s. These kinds of innovations are a mixture of business models and technological improvements, and they are included within the boundaries of this paper.

Only sheer marketing innovations as, for instance, advertisement campaigns and branding initiatives are considered outside the scope of this investigation.

A special issue in the telecom area is concerned with the distinction between product and process innovations. These distinctions cannot be sharp, as the products – the communication services – are delivered simultaneously with the process of production, so to say. Product and process can, consequently, not be clearly separated. However, not all process innovations become visible to the end user in the shape of new products/services. Moreover, it is possible to differentiate between network and service innovations, i.e. the innovation which the end users experience (services) and the innovations which end users do not necessarily notice (network).

An additional angle on the character of innovations deals with their more or less radical character. In a historical perspective, it can be useful to differentiate between technological innovations, which are within or beyond the existing technological paradigm[1]. However, in this paper regarding the innovations inside a given technological paradigm, the most relevant thing is to differentiate between incremental (step-wise) and more radical innovations. An incremental innovation is, for example, a new service developed on the basis of an IN product development platform. A radical innovation is, for instance, the implementation of 3G systems in the mobile area. The transition from a 2G to a 3G system can be made more or less smoothly. However, a fully developed 3G system will lead to radically new communication possibilities.

There are two main questions to be examined in the paper. The first one is related to the general innovativeness of the Danish telecom market as compared with other comparable markets. It has often been mentioned in the public debate in Denmark on telecommunications that the Danish market has benefited from the fact that large international operators have been able to use the Danish market as a kind of test bed because of its limited – however large enough – size and because of the general wealth and advanced demand on this market. The question, however, is whether this is true, i.e. whether innovations are implemented more quickly on the Danish market compared with other national markets in Europe. Furthermore, in relation to this issue, it will be examined how the 'innovation system' develops in Denmark. This question is, for example, concerned with the relationships and interfaces between equipment and hardware manufacturers, telecom operators and content producers and aggregators, and it, furthermore, deals with the cooperative relationships between telecom companies and universities. (In the paper, we will concentrate on the relationships between operators and equipment manufacturers).

The second main question in the paper deals with the relationships between innovation and competition, including the influence of innovations on the development of competition, but also the influence of competition on innovativeness. It is obvious that there are mutual influences between innovation and competition, where competition can increase the innovation activities of companies, while in-

[1] See, for instance, Giovanni Dosi: 'Technological Paradigms and Technological Trajectories', Research Policy 11, 1982.

novations can lead to general market developments and increasing market shares. Innovation activities can, therefore, also contribute to a sharpened competition, which does not necessarily lead to more competitors and more equal competition between operators of comparable sizes. A sharpened competition, resulting from the innovation activities of one or a number of operators, can also result in a more skewed development in market shares.

In relation to this, it should be mentioned that the theories on the positive correlation between competition and innovation does not flow from the traditional Schumpeter-based theories. They, actually, predict a negative correlation, so that an increased competition may lead to less innovation activity, as competition will diminish the monopoly rent shared by successful innovators[2]. Newer economic analyses, however, indicate positive relationships between competition and innovation. The telecom area seems to be a clear example of a sector, in which competition has led to greater innovation activity. If the period before the liberalization of the sector is compared with the present situation, it is obvious that there are more innovation initiatives in the present situation. But there is an analogy between Schumpeter's theory and the comparison between innovation activities of the dominant operator and the smaller operators on the market. Is it primarily the smaller operators or the big one which is in the lead with respect to innovation activities?

In the paper, there is first a very brief presentation of the Danish telecom market. After this, the general innovativeness on the Danish market is examined, followed by a discussion of the market based factors affecting innovations: Domestic demand, strategies of operators, competition, and finally cooperative relationships between operators and equipment manufacturers.

2 The Danish Telecom Market

The Danish market has, in the same manner as the markets in other Nordic countries, a high penetration of most kinds of telecom services. This is not a new phenomenon. Before the liberalisation processes were initiated in the mid 1990's, the Nordic countires were also characterised by high penetration rates of telecom services (tables 1 and 2). Only in Finland was the penetration rate for fixed lines below the EU(15) average, and this can partly be explained by its rapid growth in mobile communication, as Finland was the first country to experience a fall in fixed line penetration due growing use of mobile phones.

The Nordic countries were among the first to introduce mobile telephones, having at one point the highest penetration. However, their lead has decreased in the past couple of years, and a number of countries both inside and outside Europe have reached the same high level of penetration.

[2] Philippe Aghion, etc.: 'Competition and Innovation: An Inverted U Relationship', The Institute for Fiscal Studies, WP02/04, 2002.

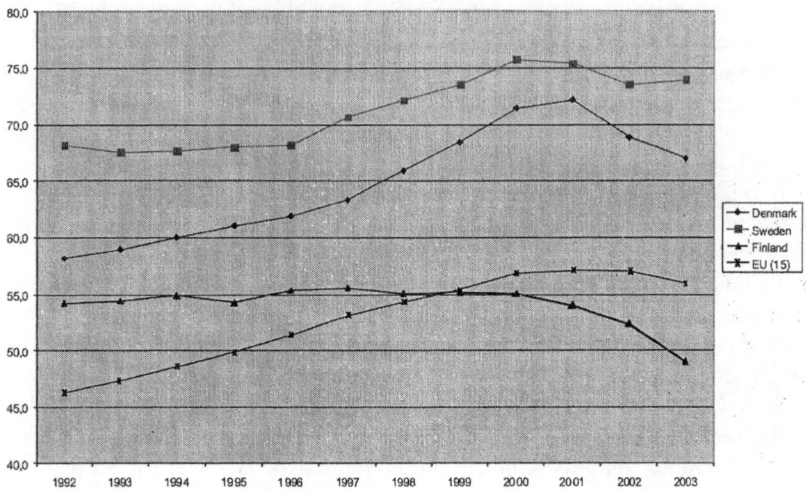

Sources: ITU database and Danish Competition Authority (2004): Telecompetition Towards a single Nordic market for telecommunication services?

Fig. 1. Penetration of fixed lines in Denmark, Sweden, Finland and EU (15

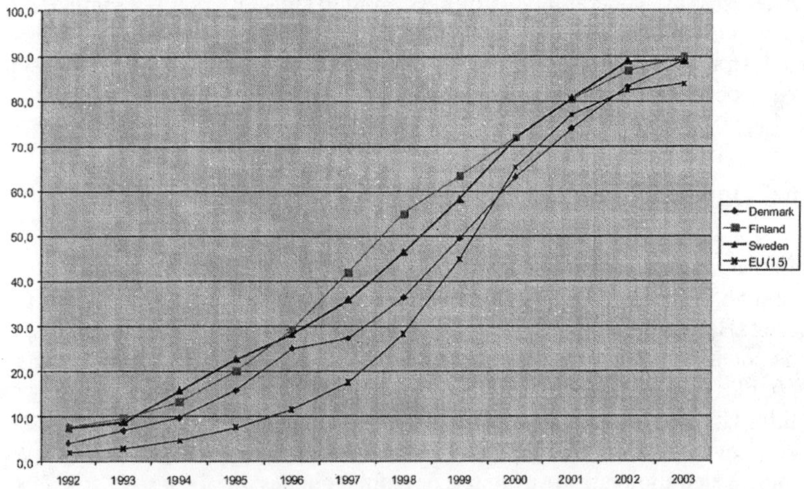

Sources: ITU database and Danish Competition Authority (2004): Telecompetition Towards a single Nordic market for telecommunication services?

Fig. 2. Penetration of mobile phones in Denmark, Sweden, Finland and EU (15)

Also with regard to Internet and broadband services, Denmark has been among the leading countries in Europe, still maintaining one of the highest penetration rates in Internet access as well as penetration of ADSL and cable modem.

Table 1. Penetration of broadband connections mid 2003 (per 100 inh.)

	ADSL	Cable modem	Others	Total
South Korea	14.3	9.3	4.3	27.9
Canada	5.9	7.0	1.4	14.3
Denmark	7.3	3.2	0.6	11.1
Iceland	9.7		0.8	10.5
Sweden	5.4	2.1	1.7	9.2
US	2.6	3.7	0.3	6.6
Norway	4.1	1.2	0.1	5.4
EU average	4.2	2.0	0.4	6.6

Source: Danish IT & Telecom Authority: Benchmark of fast access mid 2003.

The high penetration of telecom services in the Nordic countries has been explained by the fact that the major share of the population is able to afford these services. In addition, Danish consumers and enterprises are among the fastest in taking up new technology. With regard to penetration of mobile services, the Scandinavian countries have enjoyed a first mover advantage initiated through the early adoption of NMT as a common standard.

Another explanation is that telecom services used to be relatively cheap compared to other countries, in particular if compared with the use of purchasing parities. This is, however, not true in the case of broadband services. Here the prices in Scandinavia are among the highest for most types of broadband services[3].

Table 2. Market shares of major actors on the Danish Telecom market in % (1h2004)

	Fixed Subscribers	Domestic fixed traffic	Mobile subscribers	xDSL
TDC	82	61	42[1)]	76
Sonofon	2	4	21	
Orange	1		12	
Tele2	6	13		6
Debitel	3		7	
Telia	4	6	10	
Cybercity				11
Tiscali		2		4

Note: 1) Including Telmore
Source: Danish IT & Telecom Agency: Telecom statistics - first half of 2004.

TDC is, as the incumbent operator, by far the largest operator with major market shares for all major types of telecom services. The most intense competition is seen on the mobile market where TDC has a market share of 42% (32% if the low

[3] Danish IT & Telecom Authority: Benchmark of fast access mid 2003.

price operator Telmore, recently acquired by TDC, is excluded). The rest of the mobile market is shared between three operators having their own mobile network (Sonofon, Orange and Telia) and a number of service providers of which Debitel is the most important. On the market for fixed phone services, Tele2 is the most important competitor with 6% of the subscribers and 13% of the domestic traffic. On the market for ADSL, TDC has been challenged in particular by the two Internet access providers, Cybercity and Tiscali. TDC increased however their market share from 37% in 2000 to 81% in 2002 and has dominated this market ever since.

Table 3. Market concentration for fixed and mobile phone services (2h 2003)

	Fixed line subscribers		Fixed lines minutes		Mobile subscribers	
	Market share CR1	1/HHI	Market share CR1	1/HHI	Market share CR1	HHI
Denmark	84	1,42	63	2,40	33	5,09
Finland	95	1,11	38	3,21	51	2,69
Iceland	92	1,16	83	1,39	67	1,80
Norway	92	1,19	69	2,04	56	2,45
Sweden	˜100	1,13	54	3,08	43	2,78

Notes: CR1 refers to the carrier with the largest market share (usually the incumbent). HHI refers to the Hirchman-Herfingdal Index. The index is calculated by adding up the carriers markets shares. 1/HHI = n, if the market is equally shared by n operators and can be used as a parameter for the level of competition.
Source: Danish Competition Authority: Telecompetition - Towards a single Nordic market for telecommunication services?2004.

Infrastructure competition on the Danish market seems to be more developed than in other Scandinavian countries, while Finland and Sweden have more competition on the market for fixed line traffic. In a European perspective, Denmark is among the countries with most competition, only in the UK is the market shares of the incumbent smaller than in Denmark[4]. Telia has however taken over the Danish affiliate of Orange, and this will lead to a lower level of competition on the Danish market for mobile services as both companies are among the four leading mobile operators.

With regard to competition on broadband services the picture is slightly different. The market share of the incumbent is among the highest both on the market for DSL and for the market for other means (mainly cable). On the other hand, use of ULL and shared access is much more common in Denmark and Finland than in other countries, where competition mainly comes from resale. This implies that there is more infrastructure competition and less service competition in these countries compared to the rest of Europe.

[4] A. Henten, M. Falch, R. Tadayoni: New Trends in Telecommunication Innovation. Communication & Strategies, Issue 54 2nd quarter 2004. IDATE

Table 4. Market shares of incumbent operators and types DSL connections used by new entrants

	Incumbent market shares			New entrants DSL connections			
	DSL	Other means	Total	Full ULL	Shared	Bit-stream	Re-sale
Denmark	78.8	35.0	65.5	47.0	2.4	36.9	0.0
Finland	63.2	47.7	60.4	52.7	14.8	32.6	0.0
Sweden	67.6	1.4	41.3	3.5	25.4	1.8	69.4
EU (15)	73.7	7.7	56.7	18.3	11.8	25.7	44.2

Source: Connecting Europe at High Speed: National Broadband Strategies, Commission Staff Working Paper Brussels, 05/05/2004.

3 Innovation in the Danish Telecom Sector

The telecom sector is considered a high-tech area. This is supported by OECD data on R&D in telecom services as well as the telecom manufacturing industry. The radio, TV and communications industry spent 7.4% of the value of the production on R&D in 1999 (compared to 7.9% in 1991). This is in comparison to an overall spending in all industries of 2.6% on average[5]. Data on R&D in service industries are sparse, but a preliminary assessment by OECD ranks post & telecommunications as one out of three knowledge-intensive service areas (the others being finance and insurance and business services, excluding real estate).

In Denmark, data on the overall R&D activities in the telecom equipment industry are included in the data for the entire ICT industry. R&D amounts here to 5.3% of the revenue. With regard to data for telecom services, data on R&D are combined with data for the broadcasting sector. Here R&D amounts to 2.0%. It follows from table 4 that R&D in the service industry is focused on innovation and that very little research is being done.

Table 5. R&D in Danish ICT industry and ICT communication (2003)

	Research (mill. €)	Development (mill. €)	Total (mill. €)	As % of revenue	Employees (Full time)
ICT Industry	36.1	135.3	171.4	5.3	1599
ICT Communication	3.2	86.5	89.7	2.0	657

Note: ICT industry includes NACE sectors 30, 31.3, 32, 33.2 and 33.4. ICT Communication includes NACE 64.2.

Source: Dansk center for Forskningsanalyse: Danish Enterprises' R&D activities within ICT. October 2004.

[5] Figures are based on data for 12 OECD countries (OECD: 'Science, Technology and Industry Scoreboard 2003', Annex 1.1)

More detailed data on R&D can be obtained at the company level. The OECD publication 'Communications Outlook' provides R&D figures for some of the largest telecom companies. The figures in this publication indicate that R&D activities of telecom equipment is even more research intensive than if combined with radio and TV. Moreover, the figures do not show any slowdown in R&D activities as a result of the downturn in the market from 2000. In absolute terms, R&D expenditures increased from around 27 billion in 1997 to more than 38 billion USD in 2001. In 2002 and 2003, companies like Cisco, Lucent, Nortel and Alcatel experienced substantial drops in both revenues and R&D expenditures. However, they have maintained approximately the same level of R&D intensity. Siemens has maintained a relatively stable absolute level of R&D activities, and the fluctuations in R&D intensities depicted in the table 2 are the result of fluctuations in revenue.

Table 6. R&D expenditures for telecom equipment manufacturers in % of revenue

	1992	1997	1999	2001	2002	2003
Ericsson	15.7	14.5	16.1	20.1	20.9	23.0
Motorola		9.2	11.1	14.3	13.6	13.9
Cisco		12.4	13.7	17.6	18.2	20.8
Lucent		12.4	13.7	16.5	18.7	17.6
Nortel	11.0[1]	13.9	13.9	18.8	21.1	
Fujitsu		7.8	7.6	7.0	7.0	6.2
NEC	8.0	7.0	5.5	6.5	6.5	6.3
Nokia		7.8	8.9	9.6	10.2	12.8
Alcatel	12.0	8.9	9.5	11.3	13.5	
Siemens	10.1	7.6	18.8	10.1	6.7	6.8
Average of 18 companies[2]		9.5	12.4	12.9		

Notes: [1]Northern Telecom

[2] Includes, in addition to the above mentioned, Samsung, Matsushita Maconi, LG Electronics, Corning, Qualcom, 3Com and Juniper Networks.

Sources: OECD: 'Communications Outlook' 1995 & 2003. Figures from 2002 and 2003 are based on annual reports from the companies.

Table 7. R&D expenditures for telecom operators in % of revenue

	1992	1997	1999	2001	2002	2003
NTT	4.9	3.1	3.4	3.3		
Deutsche Telecom		1.8	2.0	1.9	2.1	
BT	1.8	2.0	1.6	1.7	2.0	2.0
France Telecom	4.0	3.5	2.2	1.3	1.2	1.0
AT&T	4.5 (1.9[1])	1.6	0.9	0.6	0.7	0.8

[1]Excluding equipment related innovation activities.
Sources: OECD: 'Communications Outlook' 1995 & 2003. Figures from 2002 and 2003 are based on annual reports from the companies.

R&D expenditures for telecom operators are not always market driven. Public ownership has, in a number of cases, implied that operators have had other obligations with regard to R&D than maximizing their long term profit. Many operators have been engaged in basic research, which might be profitable at either the industry or the national level, but where the immediate economic benefit for the operator was more doubtful. R&D activities have, therefore, been restructured with more focus on D than on R. OECD notes that even today operators with significant state ownership tend to spend a larger proportion of their income on R&D. In addition, a few countries including Canada, France, Japan and Korea have made specific requirements to operators with regard to their spending on R&D.

Still, some decreases can be observed in R&D for operators (table 3). Deutsche Telecom and BT have spent 1.5-2.0% on R&D throughout the period, but NTT, France Telecom and AT&T have all reduced their R&D considerably. The figures for AT&T may, however, not be as illustrative as the others, as AT&T has gone through major restructuring including divestiture of Lucent during this period.

TCD, the incumbent operator in Denmark, reports that that their R&D activities mainly are focused on product innovations, while the majority of the other operators seem to concentrate on process related activities. TDC is, by far, the Danish telecom operator with the highest R&D expenditures. Their expenditures amounted 15 million € in 2002 or around 0.7% of their domestic revenue. A major reorganisation of TDC's R&D activities has taken place since the telecom market was liberalized in the 1990s. Before the merger into TDC, the regional monopoly operators financed substantial research activities. Basic long term research was conducted by a national research centre (Tele-technical Research Laboratory), owned by the operators in cooperation, while short term product development was made in research departments within the regional companies. Following the merger of the regional companies, all research departments and the national research centre were merged into TDC research, and focus was moved towards more business oriented development activities. In 2000, TDC Research was dissolved, and innovation activities were delegated to the various business units of TDC.

At the same time, a number of new operators have entered the market. Some of these have no tradition for research, and their R&D budgets are much smaller than those of the incumbents. The overall trend in the level of innovation activities within the telecom sector depends on how such activities are defined. Most of the Danish telecom operators do not carry out their own research. On the other hand, they constantly seek to optimize their production, implement new services and create new opportunities for their business.

Development of new products is still important for the telecom operators, but product development is organised differently than before. The operators are now using much more standardized equipment, and this implies that all operators basically are able to provide the same types of services. Product innovation is, therefore, focused on adaptation of existing products to the company's own commercial and technical environment. The development within TDC is a good example of this. TDC has no separate research and development department, but maintains a large number of innovative activities in all parts of its organization. Some even claim that the overall expenditures allocated for R&D are rising although they are less visible today.

Development projects have become smaller, more commercially oriented, and with a shorter time horizon, but they have also become more numerous. Others claim that innovative activities with relation to fixed network operations are decreasing, while innovations within mobile networks and services are either stable or increasing. Cybercity, one of the major Internet access providers, claims that 20 percent of their staff are involved to some respect in innovative activities. The innovative activities include development of web-based customer handling, workflow and billing systems, development of new services and network optimization. Tiscali, the other major Internet access provider competing with TDC, has a much lower level of innovative activities, as Tiscali has concentrated most of their R&D in their Italian parent company.

On the mobile market, the second operator, Sonofon, reports that 200-250 out of a total staff of 1,300 are engaged in innovative activities. Half of these work with product development. The third mobile operator Orange (formerly Mobilix) used to devote substantial resources to R&D and was the most innovative mobile operator on the Danish market. The telecom crisis in 2001, however, forced Orange to make substantial cuts in their Danish R&D budget and is now being taken over by the Telia conglomerate in selected areas.

4 The Role of Domestic Demand

Domestic demand affects innovations in two different ways. The demand affects the speed of innovations; sufficient demand is necessary to create the economic basis for investments in new innovations. In addition, such investments will become economically attractive only if they can be expected to result in additional revenue in the future.

The demand also affects the direction of innovations. The demand for different types of services is an indicator of the types of innovations that will become the most profitable. One of the decisive factors for a company's ability to innovate is its ability through the market to receive the right signals on the future needs of its users. Therefore, the relationship between users and producers is important for a description of national systems of innovation[6]. The role of the demand is also included in Michael Porter's theory on competitive advantages[7]. Here, demand – in particular demand from advanced vanguard users – is included as one of the four factors determining competitiveness.

It follows that not only the size but also the type of demand is important for innovations. A large demand limited to a few basic services will not stimulate product innovations but may stimulate process innovations which increase capacity and reduce costs. On the other hand, even a limited demand for new services may be important for gaining experiences in product innovation. Access to advanced and demanding customers does not only contribute to the overall revenue, but this will also contribute to the learning of the company. Advanced telecom users could, for instance, be the financial sector or public institutions.

Another characteristic of demand which impacts on innovations is price sensitivity. If the demand is very sensitive to price changes, it might be difficult to introduce innovations which will improve the level of services but also result in a slight increase in costs.

Price sensitivity depends both on consumer preferences and the level of competition. Fierce price competition can, therefore, dampen innovations. Nevertheless, competition also stimulates innovations which can be used to maintain or even increase market shares. The most prominent example in Denmark of demand driven innovations within the telecom sector is the building of the partly optical hybrid cable-TV/data network in the 1980s. The building of the network stimulated demand for optical equipment and formed the basis for its current stronghold in this industry. The network did not result in the expected innovations in broadband services, but it contributed to the creation of technical competence within the area of optical communication.

A more recent example of demand led innovations is that the demand for fast and cheap Internet connections has led to the creation of alternative network infrastructures operated by municipalities, power companies or by local groups of citizens. One example is the municipality of Esbjerg, which has upgraded their cable-TV network so that they are now able to offer Internet connections at prices much lower than offered by the telecom operators.

Demand can be stimulated both through private demand and through demand from public institutions. The public sector is a comparatively advanced user of IT products and has stimulated private use of advanced IT as well, and the public sector has in this way been able to promote demand for IT- and telecom services.

---

[6] Lundvall, B-Å (ed.): 'National Systems of Innovation – Towards a Theory of Innovation and Interactive Learning', Pinter, 1992.

7 Michael Porter: 'The Competitive Advantage of Nations', The Free Press, 1990.

There are several examples in which public use has stimulated innovations in areas where demand from the private sector alone hardly would justify investments in innovations. One example is the use of EDI in the health care sector. Early implementation of EDI in public hospitals has initiated development and a wide diffusion of EDI not only in the health care sector but in other sectors of the economy as well.

Another factor in the creation of an advanced demand is the development in areas with specialized communication needs. One of the drivers in the market for Internet access and high-speed connections has been the growth in telework. Telework has been stimulated through tax deduction schemes for employer-financed home PCs. Telework has both stimulated demand for telecom connections and for various types of office products providing secure and efficient communication between the home-based work place and the office. This is an area with great potential for the development of mobile services which can contribute to demand for UMTS, Wi-Fi and other wireless services.

5 Denmark as a Test Market for Telecom Products

The concept of test market has played an important role in the Danish debate on industrial policy. The concept is used in the presentation of the Danish industrial IT policy published in 1998[8], and in later reports published by the Ministries of Research and Trade and Industry[9].

Several comparisons position Denmark as an advanced market for ICT products, in particular in the mobile areas. According to a benchmarking analysis measuring 26 parameters within infrastructure, applications and market structure, Denmark is the most advanced ICT market next to Hong Kong and ahead of other Scandinavian and European countries[10]. In a more recent benchmarking by OECD, Denmark ranks number six, while Sweden and Finland rank number one and three, respectively. With regard to private consumers use of ICT, however, the same OECD study ranks Denmark number one, while Sweden, Norway and Finland rank as number 3, 10 and 11, respectively.

---

[8] Erhvervsministeriet: 'IT/Tele/Elektronik – Danmarks ervhvervspolitiske strategi', 1998.

[9] Forskningsministeriet: 'Sund konkurrence og ægte valgfrihed', 1999 and Erhvervsfremmestyrelsen: 'IT/Kommunikation – en erhvervsanalyse', 2001.

[10] ITU, Geneva, 2002.

Table 8. Benchmarking of National ICT Markets

	ITU Mobile/Internet Index 2001	OECD ICT application benchmark
Denmark	2	5
Finland	12	3
Sweden	3	1
Norway	6	11

Sources: ITU: 'Internet for a Mobile Generation', Geneva, 2002. AH Larsen & Lotte Langkilde: Digitalisation of private businesses – a Benchmarking study . FORA October 2004.

The Danish market has in previous decades been used as a test market for electronics. IBM has, for instance, used Denmark as a test market for the introduction of personal computers. Companies wishing to be visible on a national market may prefer to start their introduction of new products on a small market.

Strategies and products can be tested before they are introduced on a larger scale. It may also be advantageous if competition with major international competitors can be avoided in the test phase. Other factors that could make Denmark attractive as a test market is that Denmark has a well-developed telecom infrastructure and, in general, a high penetration of ICT products.

In addition to this, new electronic products penetrate the market much more rapidly in Denmark than in most other European countries[11]. A survey made in 2000 among Danish companies within IT and communications confirms that Denmark was still considered an important test market in the period from 2000 to 2005 [12].

It has been argued that Denmark is well suited for a position as a test market, because the Danish telecom market is one of the most open and competitive markets within the EU. This is a result of a deliberate Government policy beginning in 1995 which introduced 'Best and Cheapest' as the overall objective for Danish telecom policy[13]. The strategy was to maintain a telecom regulation, which was ahead of the rest of the EU in terms of liberalization and creation of real competition. Early liberalization has been decisive for new operators entering the Danish telecom market and has been the major driver for the large investments made in the Danish telecom infrastructure in the mid-1990s[14]. The most obvious example of this is the entry of France Telecom through is establishment of Mobilix in 1997 (see box 1).

[11] 'The International Takeoff of New Products: The Role of Economics, Culture, and Country Innovativeness - Do new products take off at consistently different times in different countries? Why?', Gerard Tellis, 2003, 22 (2) , pp.188-208. In: Marketing Science.

[12] Ministry of Trade and Industry.

[13] Ministry of Research and Information Technology: 'Best and Cheapest by Way of Real Competition', Copenhagen, 1995.

[14] See, e.g., Andersen Management International: 'Konkurrencesituationen i telesektoren i Danmark' (The level of Competition in the Danish Telecom Sector), report to Ministry of IT and Information Technology, Copenhagen 1999.

Box 1 Mobilix/Orange

Mobilix was established in Denmark in 1997, and it won one of the four GSM1800 licenses which were offered this year. From the beginning, the ambitions went beyond the Danish borders. Mobilix was one of France Telecom's first major engagements in a foreign national market in Europe, and it was planned for Mobilix to cover the Scandinavian market, which at that time was the most advanced in the world with respect to mobile services.

Denmark was seen as a market suitable for development and testing of new services. From the outset, substantial investments were made in marketing, infrastructure and development of a new organisation.

From March 1998, Mobilix offered national coverage for mobile telephony. As of January 1999, Mobilix offered fixed services as well, and in October 1999 Mobilix became a full-scale telecom operator offering all types of telecom services. In these years, Mobilix was recognized as one of the most innovative operators on the Danish market, introducing a number of new services. In cooperation with a national clearing bank, Mobilix developed one of the first mobile payment systems using the mobile phone as a personal terminal for electronic payments with debit cards.

The investments in the Danish market were not matched by the immediate revenue obtained, and the Danish affiliate generated heavy deficits. In December 2000, Mobilix became an affiliate of Orange, and in particular after the international telecom crash, major cuts in the budget were demanded. Innovation activities were substantially reduced and, presently, product development is primarily made by the parent Orange company. The Danish division of Orange was, however, able to maintain a R&D department with 20-30 employees doing R&D for all of Orange until 2003, as Orange Denmark had acquired special competencies in certain areas. In 2004, Orange Denmark was sold to Telia.

Sources: Interviews and press releases from Orange/Mobilix

The Danish telecom market is still among the most advanced in Europe, but Denmark has not been able to maintain its position as a test market. Sometimes, the same product is tested in a number of countries using different pricing models or marketing strategies in each country, and Denmark does not have a unique position in that respect.

Products introduced by Sonofon are often first tested on the Norwegian market by Telenor (the majority share holder of Sonofon). Nor does TDC see Denmark as a particular test market. When operators are asked to compare Danish and foreign markets, they do not see the Danish market as particularly innovative. They rank the innovativeness of the Danish market for fixed network services as being on the average level of the EU countries, while the mobile market is above average but far from a leading position. In particular, the markets of the UK and Norway are more advanced than the Danish, and also the Swedish market is more advanced in certain areas.

Denmark has not been at the fore in the development of UMTS, as it was in the development of GSM and especially NMT. Even though 3G services already have been available for a couple of years in both Japan and South Korea and in a number of European countries, such services are only offered in Denmark by 3 – the newcomer on the Danish 3G market – since the end of 2003, while the other 3G licence holders including TDC do not yet provide 3G services. A reason for this could be that Denmark was among the last countries within the EU to auction its

UMTS licenses. Another reason is that the Danish licenses are stricter with regard to network sharing between the licensees. This increases the level of investments in network infrastructure needed to offer 3G services and may delay its introduction. On the other hand, it will in the long term enable more competition in the development of new networks and services.

More serious, however, is that the Danish telecom industry has been unable thus far to make the necessary investments in order to build up core competencies within 3G at an international level. UMTS is a much more complicated technology that GSM and NMT. It is, therefore, more difficult for small enterprises to afford the investment needed to stay competitive, particularly in a situation where the Danish market is less advanced than in a number of other countries.

I-mode is another technology that not has reached the Danish market. When NTT DoCoMo wanted to test this technology in Europe, they chose to start off in Germany and the Netherlands. The Danish market did not seem to be attractive, maybe because of the fierce competition on prices.

Responses to a questionnaire distributed to national regulatory agencies in selected countries reveals that the timing for introduction is very similar on the different markets. It is not possible to find a clear trend, but Germany has been early in their adoption of all services included in the questionnaire. None of the services listed seem to have used Denmark as a test country (maybe apart from GSM, but that cannot be confirmed by the questionnaire). It is important, however, to note that the time of introduction is determined by a combination of market and regulatory conditions, as most of the services demand a license before they can be introduced. A few services not included in the questionnaire have been introduced in the Danish market first, for example Multiplan (first developed and marketed by TDC under the name Duét) and mobile payment (introduced by Orange). The Danish position as a test market has been lost during the international telecom crisis. Focus was moved from innovation to cost reduction and price competition. Additionally, the markets for mobile communication have matured, and this has led the operators to concentrate on the markets with the largest volume. Finally, a delay in auctioning of UMTS licenses has prevented a vanguard position on the 3G market at least in the short run.

Table 9. Time of Introduction for selected services

	GSM	GPRS	ADSL	FWA
Denmark	1992	01/2001	07/1999	12/2000
Sweden	1992	10/2001	Spring 1999	12/2001
Norway	1993	1/2001	6/1999	3/2000
UK	1992	4/2001	12/1998	9/2000
Netherlands	1994	12/2000	11/1999	2002
Germany	1992	12/2000	1/1999	8/1999
France	1992	5/2002	11/1999	No information

Source: Responses from national regulators

6 The Role of Competition

This section is concerned with the relations between competition and innovation with respect to the implications of competition on innovation activities, on the one hand and the implications of innovations on the competitive environment on the other. These two questions are often seen as two sides of the same coin, as companies innovate to have a stronger position in the competitive environment and that competition, consequently, increases innovation activities. However, it is not always that simple, and it is important to examine the two questions more thoroughly.

The implications of competition include the following points:

- The extent of innovation activities
- The innovators (who is doing the innovation)
- The character of innovations

The implications of innovations include the following points:

- Market shares
- The character of competition

6.1 The Implications of Competition on Innovation Activities

A comparison of the extent of innovations in the telecom sectors before and after the processes of liberalisation speaks in favor of the competitive market situation. However, the supposition that competition leads to increased innovation activities does not apply universally in all situations. In the Schumpeter-based innovation theory, on the contrary, it is argued that competition can hamper innovation, as there is less economic room for innovative activities[15]. In the telecom sector, however, it has clearly been proven that innovation activities have grown considerably with the introduction of liberalisation. This does not imply that the sector formerly was lacking innovativeness; many of the technologies which, at present, are implemented have been developed in the monopoly period. Yet the incremental innovations are currently far more numerous, and the degree of variation is considerably greater. A radical innovation such as the Internet which, so to say, comes from outside the traditional telecom sector would not have developed in the same manner and seen such a wide diffusion without a liberalisation of alternative operators' access to the telecom networks.

The question of who the most innovate operators are can be analyzed along the lines of the above discussion on whether competition enhances or limits innovativeness. Is it primarily the big operator, TDC, or is it first and foremost the newcomers which innovate? The answer is far from simple and needs clarification. A

[15] This theory has found some empirical evidence – see, for instance, Luc Soete: 'Firm Size and Inventive Activity: The Evidence Reconsidered', European Economic Review 12, 1979.

company like Cybercity has been and is continuously an innovative force in the Danish telecom market. The company was, for example, the first to introduce ADSL in Denmark. Telia was likewise in the beginning very active on the Danish market with considerable investments, but has since then had to limit its investment and development activities because of great financial problems. Telia has cut its investment activities in Denmark down to a level of 20-25% of what it was in 2000, and the development staff has been reduced to one-third[16]. The reason for this downscaling of activities is that in many market segments, Telia has not seen measurable success in the Danish market. In a start-up phase, it is possible to 'invest in the future', but in the long run it is not possible to keep on investing if this does not result in a positive economic outcome, says Telia.

TDC, the incumbent on the Danish market, has initiated a number of market introductions during the past decade. The most often mentioned innovation is the Duét service, which integrates mobile and fixed network telephony for the customers, and which, for example, won the innovation price in 1999 at the big ITU technology exposition in Geneva. However, TDC is often not the first to introduce new services on the Danish market. In most cases, TDC follows the first innovator but eventually manages to become the biggest player in the new market segment. This applies to some extent to Internet access – the so-called ISP area – and it applies to a large degree to ADSL. With respect to cable modem service, Telia Stofa was the first operator on the market. However, the market force of TDC is larger, as TDC has more cable-TV customers than Telia Stofa. An overview of different innovations and the innovating companies is provided in table 9.

Table 10. Who was first? Examples of services and technological solutions and the first innovators

	First innovatør
IN-based services	TDC
Internet acces	UNI-C (UNI2)
Duét	TDC
Taletidskort	Sonofon
National mobil roaming	Sonofon
GPRS	Sonofon
Location based services	Orange
Mobile payment	Orange
MMS	TDC
ADSL	Cybercity
Cable modem	Telia Stofa
WLAN	non-operators

16 Information from interview with Ole Krogh Buus, Telia.

An often given explanation for the relatively late market introduction by the incumbent operator is that the incumbent has to earn back its considerable investments in earlier technological solutions, for instance ISDN in relation to ADSL, or that the incumbent already has technological and economically profitable solutions in its product portfolio and does not immediately see any reason to supplement with other solutions, for example leased lines in relation to SDSL. The explanation is thus that an operator, which already has a large market share on a given market, does not want to 'cannibalise' its own market which follows from general business logic. There will, however, continuously be an evaluation of when it pays to offer substituting or supplementing products.

The explanation as to why the big operator often manages to become dominant also in new market segments is – in contrast to the theory predicting that first-comers will win the market – that the big operators mostly have a solid technology competence and financial capacity, but also that it is possible for them to build on the existing dominance in complementary markets. New technological solutions mostly build on the existing infrastructure. ADSL is a much discussed example in this respect. TDC was not the first with ADSL in Denmark, and Cybercity and Tiscali rapidly acquired large market shares of the small ADSL market in the beginning. Yet after a couple of years, TDC came to dominate the market with almost 80% of all ADSL subscribers. This has resulted in accusations against TDC regarding price dumping and delays in the procedures with respect to service delivery to customers of other operators. Both accusations have, however, been rejected by the competition authority and the IT and telecom agency. However, the accusations have not been given up by the plaintiffs, and the complaints continue in the system.

However, no matter what the end result will be, it is not strange from a general economics point of view that there are transaction procedures and, therefore, costs incurred in relation to interactions between different operators. The operator which owns the network will generally have an advantage over other operators, and the role of regulation will be to diminish all transaction procedures and costs for alternative operators as soon as they arise. The problem, in an innovation context, is that this can limit some of the incentives for the company owning the network to develop new applications if equal opportunities are to be given to other operators at once. This, however, may be necessary in a situation where one operator possesses the overwhelming part of the access capacity.

The conclusion with respect to innovativeness is that one cannot unequivocally determine whether the incumbent operator or the newcomers are the most innovative. If one concentrates on products which substitute for existing services (for example, SDSL instead of leased lines) and where there may be a certain cannibalization of own products, there will be a tendency for new players to come first to the market with new services for the customers. If, on the other hand, we are dealing with products covering new demands (for instance integration of fixed and mobile networks), it can just as well be the old operator which initiates innovation. Furthermore, if one considers the big financial problems of the new operators, limiting their innovativeness and pushing them towards building a profitable business on their existing product portfolio, then the incumbent operator will to an increasing degree dominate innovation activities.

This may have implications for the total amount of innovations in the market. If a competition mainly based on price competition becomes too intense, it may have negative effects on the degree of innovativeness. This situation exists in Denmark according to Telia. They believe that there is far too much weight in the Danish telecom policy under the heading of 'best and cheapest by way of competition'[17] on cheapest and too little weight on best (including innovativeness). With respect to the character of innovations as a consequence of competition, there are three important issues, which are dealt with here:

1. the increasingly demand-driven nature of innovations
2. the dominance of service developments as compared with genuine research activities
3. the internationalisation of the telecom sector

The two first questions are intimately interrelated. In a competition-oriented market, the players will follow a logic according to which innovations are guided by demand to a larger extent than in a monopoly market. In a monopolistic situation, the operator can introduce new services where and when it fits into an overall plan, including depreciations of innovation activities. In a more competitive situation, the operators are more inclined to follow the logic of the market.

This has led to a situation in which the activities of innovation on the part of operators must currently be characterized as service development rather than genuine research activity. Genuine research activities occur mainly within the equipment manufacturers (and the universities), meaning that differentiation in the basic network and service deliveries, in principle, are rather small among the different operators. The operators can buy the same basic elements from the equipment producers. In addition, the emphasis in the innovation activities of the players in the telecom transport market will be on service development, with linking and interoperability between functionalities and billing innovations playing a crucial role. This division of labor, which is increased by competition and which pushes the operators towards a concentration on their core activities (i.e., to deliver communication and information services to their customers leaving the genuine research activities to the equipment manufacturers) has the effect that the tendency towards product differentiation which competition induces will primarily be expressed on the service front, especially with respect to the smaller and incremental service elements. The problem with such incremental innovations is that they can be relatively easily copied and that a sharp price competition may be the result. In the last instance, it may therefore turn out that the competitive struggle will depend mostly on the point of departure of the operators with respect to network infrastructure and financial capacity with respect to renewal and extension if their network.

---

[17] This was the title of the telecom policy of the Danish government published in 1995 and remains the main focus of telecom policy developments. (Ministry of Research: 'Best and Cheapest through real competition', 1995).

Liberalisation and internationalisation of telecom markets have led to internationalisation and commercialization of innovations made by the operators. It could, therefore, be argued that the term 'national systems of innovation' is becoming less meaningful, as innovations are made by international companies and not by national companies. The example of Tiscali could indicate that innovations on the Danish telecom market depend more on its competency in the Italian telecom sector than in the Danish market. The creation of more multinational telecom operators may point to a concentration of innovation activities within countries hosting parent companies, while innovation activities in affiliates are reduced. However, innovation activities may also be concentrated in those markets with the highest level of competence in a particular field. The Danish market reveals examples of both trends. The Danish affiliate of Tiscali pursues hardly any research or development, while the Danish affiliate of Orange has carried out research and development for all of Orange in certain areas. In the same manner, TDC has spread their innovation activities not only among several locations in Denmark but also among foreign affiliates. The organisation of R&D in both Orange and TDC thus provides examples where the local availability of competence seems to affect the location of activities.

6.2 The Implications of Innovation on Competition

The question of how innovations affect competition – including the question of whether an innovative company will win market shares from companies with less innovation activities – is partly dealt with in the above sections in relation to the ADSL development. However, before focusing on market shares it should be mentioned that the new services and communication possibilities result in total market growth from which all market actors will benefit. It is thus not a static market situation with a stable demand to be distributed in different ways among the operators. This was, to some extent, the case earlier on whereby the share of income used by a family for telecommunications remained very constant. However, this does not apply any more in a situation in which mobile communications and the Internet have been added on top of the telecommunication consumption of families. In like manner, it does not apply to the telecom usage of businesses, which has increased considerably over the past years.

These issues also have relevance in relation to the interoperability between new innovative services. The individual operator has an interest in increasing its turnover, but there can be a trade-off regarding the interests in either a smaller share of a big market or a large share in a small market. The big market can be obtained if there is agreement between the operators on standards and accounting principles, while a smaller market may be the result of a lack of agreement. This problem is relevant to MMS, where the operators have had a hard time agreeing on the accounting principles, a fact which has held back market development in this area. The incentive to buy a mobile terminal supporting MMS and to use MMS is limited if MMS messages can only be sent internally among the subscribers of the operators separately.

On the whole, it is difficult to unequivocally conclude that innovativeness leads to increasing market shares. Many other factors also play a role, first and foremost price competition and marketing campaigns. Yet the cases mentioned below illustrate that innovative operators in different situations have won considerable market shares as a consequence of innovation activities. As already mentioned, this does not necessarily mean that the first innovator will keep its lead. Small innovative companies can quickly be overtaken by other telecom companies, as the ASDSL example illustrates.

The DSL technologies were developed more than 10 years ago but were not introduced on the Danish market until 1999 with the introduction of ADSL. Cybercity and shortly thereafter Tiscali (Worldonline) were the first operators on the ADSL market, and they quickly acquired large market shares of the initially small ADSL market. In 2000, Cybercity and Worldonline as well as TDC had each approximately one-third of the market. This does not correspond to the general development in Europe where it was primarily the old former monopoly operators which from the beginning dominated this market. However, the 'raw cupper' was opened for access for other market actors already in 1998 in Denmark, two years ahead of EU decisions in the area. This has not only brought about another distribution of market shares in Denmark but also a higher penetration. On the basis of the new technological possibilities for offering high-speed access to Internet and because of the competition between Cybercity and Tiscali – and maybe also because of the fact that TDC did not get a FWA license in 2000 – TDC opted for at speedy extension of the ADSL access availability. At present, almost 97% of TDC's customers have access to ADSL. With respect to market shares, TDC has become very dominant on the ADSL market and has more than 80% of the total number of ADSL subscribers, while Cybercity is down to 10% and Tiscali down to 6% (see table 2). It has been difficult for Cybercity and Tiscali to retain their relatively good point of departure because of the complementarity between ADSL and the physical access network owned by TDC.

Cable modem service is one alternative to ADSL in high-speed Internet access. Cable modem service was introduced in 1998/99 in Denmark. StofaNet was the first provider, but since then TDC has also arrived on the scene. StofaNet has given Telia a good possibility for offering high-speed access, while for TDC cable modem service is more of a supplement to ADSL. As a 'first-mover', Telia now holds the largest market share of the 133,000 cable modem customers in 2002. Telia has two-thirds of these customers and TDC has one-third. However in a longer term perspective, it is very likely that TDC will be the biggest operator in this market segment as well, as TDC cable-TV has about 800,000 customers, while StofaNet has only 350,000. TDC's potential is thus by far the biggest.

Duét is a service which combines mobile and fixed access to customers. The Duét service was launched by TDC in 1997/98. From the beginning, Duét met with complaints from Sonofon regarding cross-subsidization (from the fixed network to mobile). However, the complaint was rejected by the competition authority which reasoned that this was a new service and in itself a profitable service. Since then, Sonofon has marketed its own combination services, DuoFon and later Multiplan, which is specifically directed however at the business customer segment, while Duét has a broader customer target. It is difficult to document that Duét led to a jump in market shares for TDC, as the service was launched in a period where competition with new operators Mobilix/Orange and Telia increased for other reasons. However, Duét must be considered a solid success and has contributed to the consolidation of TDC's position in the mobile market.

Prepaid mobile cards are incremental innovations. They were introduced on the Danish market in 1998/99 but cannot be seen as a clear success with respect to the development of market shares in this period. It was in the same period that Mobilix/Orange and Telia had their entry in the market, and both of these operators (especially Mobilix/Orange) quickly acquired considerable market shares. The effect of the prepaid cards from Sonofon can thus best be measured by the limitations in losses of ist market shares. Moreover, prepaid cards are a type of innovation which other operators can similarly offer relatively quickly, which also happened.

The gains from Duét and prepaid cards are thus hidden in the total market growth and in the market shares of the newcomers, but it can still be concluded that there are market shares to be gained by being the first innovator on the market. However, whether this gain can be sustained depends on many circumstances, including strengths in complementary areas (ADSL), development potentials (cable modem), new substituting products (Duét) and price competition (prepaid cards). Price competition is especially a serious problem for smaller incremental innovations which can be copied relatively easily. This point is emphasized by Sonofon[18], where the analysis is that the advantages offered by product innovations can easily be overtaken and exposed to price competition, while process innovations are often of a more fundamental character resulting in productivity gains, which can have a more long-lasting effect on competition. Product innovations will, therefore, often lead to short market advantages, while process innovations can provide more long-term competitive advantages and increased market shares. Sonofon claims that the convergence between mobile and fixed networks is an area in which process innovations may provide more long-lasting competitive advantages. In addition, Next Generation Network developments also represent an area in which operators can gain more long term market advantages.

Finally, with respect to the question of whether innovation leads to more competition, this depends partly on how increasing or diminishing competition is defined. If increasing competition is defined as a growing number of operators and a more equal distribution of market shares, it is not given that innovation leads to increasing competition. This applies, for instance, if the incumbent operator is the primary innovator. If increasing competition is defined as a sharpened struggle between the operators, however, it can positively be confirmed that innovation leads to increasing competition. Under all circumstances, it leads to another and more dynamic form of competition as compared to the sheer price competition or a competition based on marketing campaigns.

[18] Interview with Kim Wehrs from Sonofon.

7 Empirical Evidence

A substantive number of empirical studies have been made in order to analyse the impact of competition on innovations. An OECD working paper by Sanghoon Ahn from 2002[19] includes a survey of such studies and concludes, among others, that:

1. The thesis that market concentration is conducive to innovation cannot be supported by empirical findings.
2. A positive relationship between product market competition and productivity growth is verified through a large number of studies.
3. Policy changes (e.g., regulatory reforms and liberalization) bring about productivity gains and long-run economic growth.

The studies on which these conclusions are based include 12 studies of the telecom sector. One of these (Gort and Song from1999)[20] is based on data from AT&T Long Lines and 8 regional companies from 1952 to 1992. This study confirms a faster development in efficiency in competitive markets than for the local monopolies. Most of the other studies conclude that competition leads to either lower prices or higher efficiency. This conclusion is quite obvious, but does not indicate anything about the impact on innovation. Only one study by Cuilenburg and Slaa addresses innovations explicitly[21]. This study confirms a positive correlation between competition and innovation in the telecom sector. Neverthless, it also shows that innovations are, to a greater extent, determined by telecom revenue per capita than by competition itself. Although these studies contribute to the general debate on whether competition induces innovation, they are only partly applicable for an analysis of the impact of the present liberalization of the telecom market, which was hardly initiated in 1992. The reservations presented by Cuilenburg and Slaa on the impact of revenue per capita is, however, interesting. At present, there is also a clear tendency toward more liberalisation in the most advanced markets. This implies that even though the most competitive markets may be the most innovative, this does not prove that more competition leads to innovation. It may, in fact, be the other way round, that innovations have contributed to a more heavy pressure towards more liberalised markets.

A more recent study[22] relates the development in investments to the level of competition by use of the reciprocal value of the Herfindahl-Hirschman Index (1/HHI) as a proxi gauge for the effective number of suppliers, using this to construct a number of indexes for the level of competition in each country every year.

The indexes are based on market data from the EU implementation reports. As these reports only provide data on market shares for the largest operator (the two largest on the mobile market), we have to use a very crude version of the HHI, ac-

[19] Sanghoon Ahn: Competition, Innovation and Productivity Growth: A Review of Theory and Evidence, OECD Economics Department Working Papers no. 317, OECD, Paris.

[20] See Sanghoon Ahn: Op.cit.

[21] Cuilenburg and Slaa: Op.cit.

[22] Anders Henten, Morten Falch & Reza Tadayoni: Op.cit

cording to which only the square of the market share for the largest one or two operators is taken into account. Although this is clearly a technical limitation, inclusion of figures from all the smaller operators would probably not reveal very different results. Indexes are constructed for the following markets: Local calls, long distance calls, international calls, mobile communication and DSL. In addition, composite indexes are made for fixed telephony (average of local, long distance and international markets), mobile and fixed (average of fixed telephony and mobile) and all combined (average of fixed telephony, mobile and DSL). The all-combined indexes are compared with investment levels in figure 3.

It follows from the figure that the countries with the highest level of investment are also those with the highest composite competition index. Yet there are a number of exceptions, for instance Finland, which has had a high growth in investments but a fairly low level of competition. The correlation between investments and the all-combined index is depicted in figure 3. The correlation coefficient between the two is 0.63.

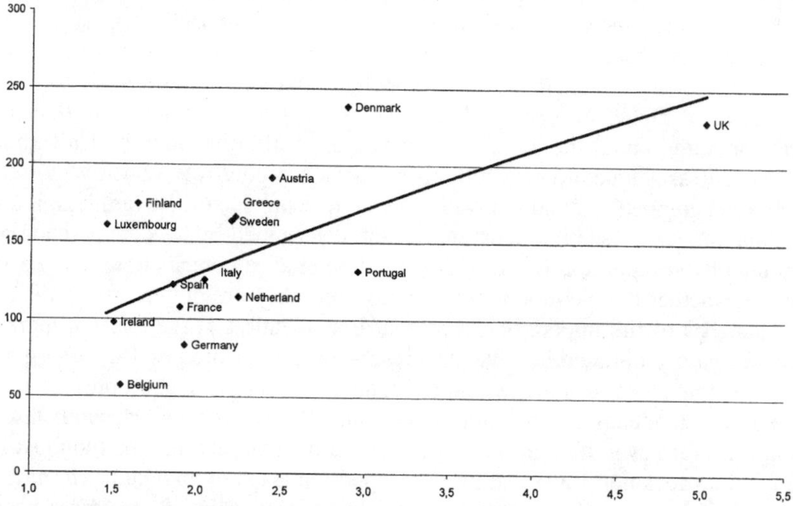

Source: Anders Henten, Morten Falch & Reza Tadayoni: New Trends in Telecommunication Innovation. Communication & Strategies Issue 54. 2nd quarter 2004.

Fig. 3. Correlation between the all combined competition index and investments per capita (2001)

Although there is a clear correlation between investments per capita and the all-combined index, it is revealing to also compare the correlation between investments and some of the other competition indexes.

8 Conclusions

The paper examines innovation activities in the Danish telecom sector. Despite the importance of innovation (also in this sector), not much has been written on the topic of telecom innovation. However, much has been written on the penetration (diffusion) of new services and on the price developments for telecom services. In like manner, much has also been published on the importance of new telecom services for the renewal of processes and products of other sectors. On the topic of innovations in the telecom sector itself, however, the amount of published material is rather scanty.

This paper concentrates on the market-based factors affecting innovation and on the implications of innovations on competition. Policy intervention is also important, as are the existing technology platforms, with their impact on path dependencies. However, the focus here is on the market related factors, including domestic demand, strategies of operators with respect to test markets, competition in the sector, and the divisions of labor between telecom operators and equipment manufacturers.

The paper emphasizes the importance of advanced demand. In Denmark, telecom demand is relatively advanced and the take-up of new services relatively rapid in international comparison. However, markets in Europe have converged to a considerable degree during the past 5-10 years with respect to penetration of services and regulatory conditions, and differences have become more limited. A result of this (in combination with the changing focus of operators resulting from the telecom crisis) has been that the test market status that Denmark clearly had in the end of the 1990s has to some extent evaporated.

The liberalization of the telecom sector has clearly changed the patterns of innovation in this sector. Telecom operators have moved towards service developments close to the market, while genuine research and development activities increasingly are taken care of by the equipment manufacturers. This is probably the clearest conclusion that one can draw from the relationship between competition and innovation in the sector.

It is a general trend in the whole telecom sector even though the Japanese example illustrates that there are other possible options for operators to play a more central role in innovation activities. It can also be concluded that competition has unleashed many innovations in the sector. Whether the innovative accomplishments are greater under a competition regime than under a monopoly regime cannot be determined from the documentation in this paper. This would require a deeper analysis of the amount and character of innovations under different market conditions. However, there is no doubt that competition has opened the way for many incremental, and also for some more radical, innovations.

With respect to the implications of innovations on the competitive environment, this paper demonstrates that innovations may contribute to increases in market shares for the innovative operators. However, there are many other factors to be taken into consideration, and often the innovations of smaller newcomers to the market are overtaken by incumbents which gain market strength on the basis of their existing market dominance.

In crude theories on competition, it is often assumed that competition will automatically lead to more innovations, as operators will innovate in order to expand or consolidate their market positions. However, reality is often more complicated than this partly tautological argument. Competition may also turn out to limit innovation if it becomes overwhelmingly price-based and if operators run into fierce financial problems. The character of competition is therefore important to include in the analysis and can furthermore be dealt with by means of policy intervention, favoring one kind of competition over the other. In Denmark, the debate on this theme is gaining weight, illustrated, by example through the inclusion of the innovation question in the latest competition analysis by the Danish IT and telecom agency. The innovation theme will probably gain more prominence in the discussions on the development of the telecom sector in the immediate years to come.

References

AGHION, P. et al. (2002), 'Competition and Innovation: An Inverted U Relationship', The Institute for Fiscal Studies, WP02/04.

AHN, S. (2002), Competition, Innovation and Productivity Growth: A Review of Theory and Evidence, OECD Economics Department Working Papers no. 317, OECD, Paris.

ANDERSEN MANAGEMENT INTERNATIONAL. (1999), 'Konkurrencesituationen i telesektoren I Danmark' rapport til Forskningsministeriet.

CEC (2004), Connecting Europe at High Speed: National Broadband Strategies, Commission Staff Working Paper Brussels, 05/05/2004.

CUILENBURG, J. and SLAA, P. (1995), Competition and Innovation in Telecommunications, *Telecommunications Policy*, vol. 19, no. 9, pp. 647-663.

DANISH COMPETITON AUTHORITY (2004), Telecompetition Towards a single Nordic market for telecommunication services?.

DANISH IT & TELEKOM AUTHORITY (2003), Benchmark of fast access mid.

DANISH IT & TELEKOM AUTHORITY (2004), Telecom statistics - first half of 2004.

DANSK CENTER FOR FORSKNINGSANALYSE (2004), Danish Enterprises' R&D activities within ICT.

DOSI, G. (1982), 'Technological Paradigms and Technological Trajectories', Research Policy 11, 1982.

DUTTA, S. & JAIN, A. (2002), 'The Networked Readiness of Nations', INSEAD.

ERHVERVSFREMMESTYRELSEN (2001), 'IT/Kommunikation – en erhvervsanalyse'.

ERHVERVSMINISTERIET (1998), 'IT/Tele/Elektronik – Danmarks ervhvervspolitiske strategi'.

FORSKNINGSMINISTERIET (1995), 'Bedst og billigst gennem reel konkurrence'.

FORSKNINGSMINISTERIET (1999), 'Sund konkurrence og ægte valgfrihed'.

ITU (2002), 'Internet for a Mobile Generation'.

LUNDVALL, B.-Å. (ed.) (1992), 'National Systems of Innovation – Towards a Theory of Innovation and Interactive Learning', Pinter.

HENTEN, A., FALCH, M., & TADAYONI, R. (2004), New Trends in Telecommunication Innovation. Communication & Strategies Issue 54. 2nd quarter 2004.

AH LARSEN & LANGKILDE, L., Digitalisation of private businesses – a Benchmarking study . FORA October 2004.OECD: 'Communications Outlook' 1995 & 2003.

OECD (2003), 'Science, Technology and Industry Scoreboard 2003'.

PORTER, M. (1990), 'The Competitive Advantage of Nations', The Free Press.

SCHUMPETER, J. (1939), 'The Theory of Economic Development', 1912. Schumpeter, Joseph: 'Business Cycles', 1939.

SOETE, L. (1979), 'Firm Size and Inventive Activity: The Evidence Reconsidered', European Economic Review 12, 1979.

Regulations of Telecommunications and Market Dynamics in Germany

Friedhelm Dommermuth[1]

1 Development since the Beginning of Liberalisation..119
1.1 Employment Development..119
1.2 Turnover Development...120
1.3 Growth in Telephone Channels..120
1.4 Broadband Accesses...121
1.5 DSL Traffic Trends...122
1.6 Access to Deutsche Telekom AG's Subscriber Accesses..........................123
2 The New Telecommunications Act (TKG) and its Impact on Future Market Trends..126
2.1 Consultation and Consolidation Procedure...126
2.2 Consistency Requirement...127
3 Conclusions...128

1 Development since the Beginning of Liberalisation

1.1 Employment Development

Compared with the previous year, at the end of 2003 the number of employees in the telecommunications services sector had fallen by 2.2 per cent (5,000) to 226,400. Compared with 2002 the drop in employment was thus lowered by nearly two percentage points, viz. more than 4,000 employees. Decisive for the drop in employees was the reduction in DT's employees in Germany by 2.5 per cent (4,500) to 173,300. The competitors, on the other hand, have largely kept their employee number constant at 53,100 compared with the previous year.

[1] The explanations given reflect the author's personal opinion.

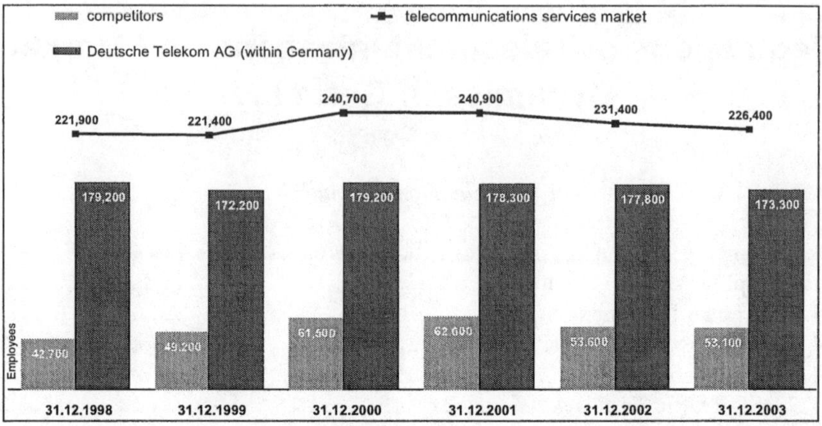

Fig. 1. Job development on the telecommunications services market

1.2 Turnover Development

The turnover volume of the telecommunications services market in 2003 totalled approximately € 63.4 billion which is an increase of about three per cent compared with the preceding year.

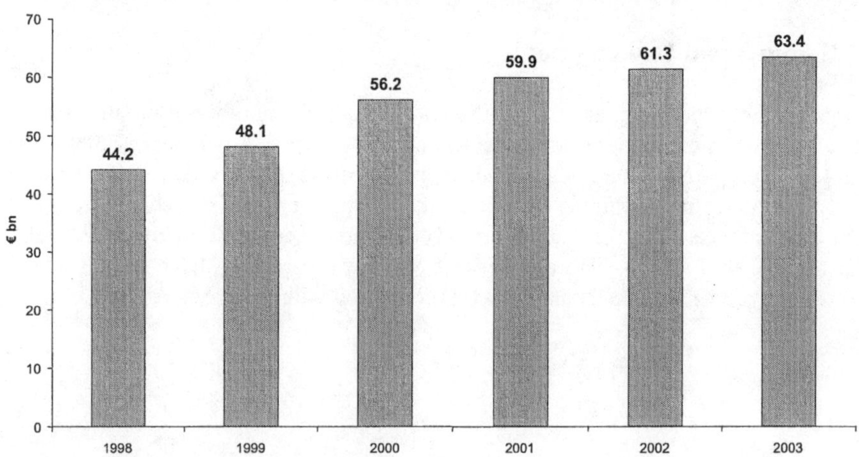

Fig. 2. Companies' turnover 1998 - 2003 from telecommunications services

1.3 Growth in Telephone Channels

At the end of 2003 there were a total of 54.35 million telephone channels in Germany. These derived from 27.8 million analogue accesses, including public telephones, 11.43 million ISDN basic accesses and 123,300 ISDN primary accesses. The competitors were able to increase their telephone channel base to 3.12 million channels in 2003. As such they achieved a share of 5.7 per cent of the total number of channels. Their nationwide share of analogue channels totalled 1.1 per cent, their ISDN basic rate share 9.0 per cent and their primary rate access share 20.8 per cent.

Table 1. Telephone accesses and competitors‘ shares

	2002		2003	
	total / m	Competitors‘ share	total / m	Competitors‘ share
analogue accesses (excl. public tel.)	29.13	0.9 %	27.69	1.1 %
ISDN basic rate accesses	10.43	6.7 %	11.43	9.0 %
ISDN primary rate accesses	0.123	17.1 %	0.123	20.8 %
public telephones	0.110	3.7	0.107	3.4 %

1.4 Broadband Accesses

Broadband accesses with transmission rates of more than 128 Kbit/s are offered over digital subscriber lines (DSL), cable television accesses (cable TV), powerlines and satellite. At the end of 2003 more than 4.6 million broadband Internet accesses were in operation in Germany of which 4.1 million were TDSL accesses of Deutsche Telekom AG, about 400,000 DSL accesses of fixed-line competitors, more than 60,000 bidirectional cable accesses, 8,000 powerline accesses and about 45,000 Internet accesses via satellite. This means that at the end of 2003 Deutsche Telekom AG's competitors had a broadband access share of about 11 per cent compared with roughly 8 per cent in the year 2002.

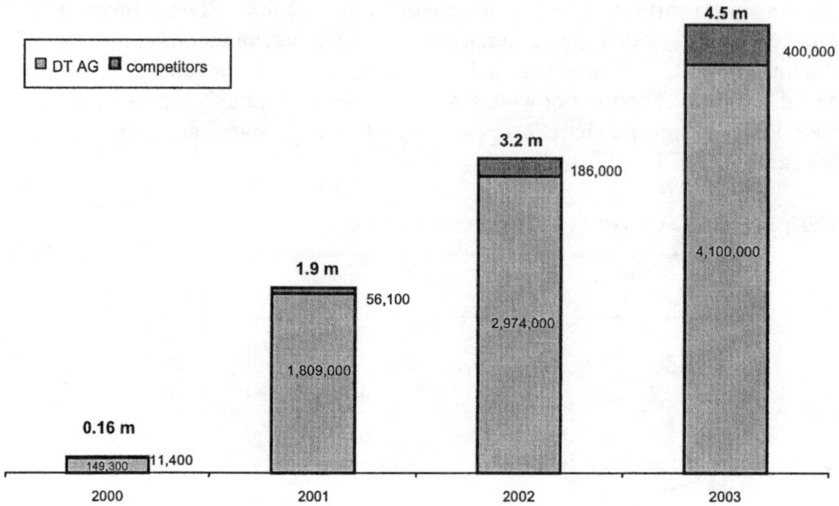

Fig. 3. Switched DSL accesses

1.5 DSL Traffic Trends

Meanwhile, narrowband Internet traffic has decreased in Germany. This is due to the fact that many high-volume users have switched from narrowband accesses to high-bit rate accesses, notably DSL. Hence DSL traffic (given the strong growth of DSL accesses) increased considerably in 2003 as compared to the previous year.

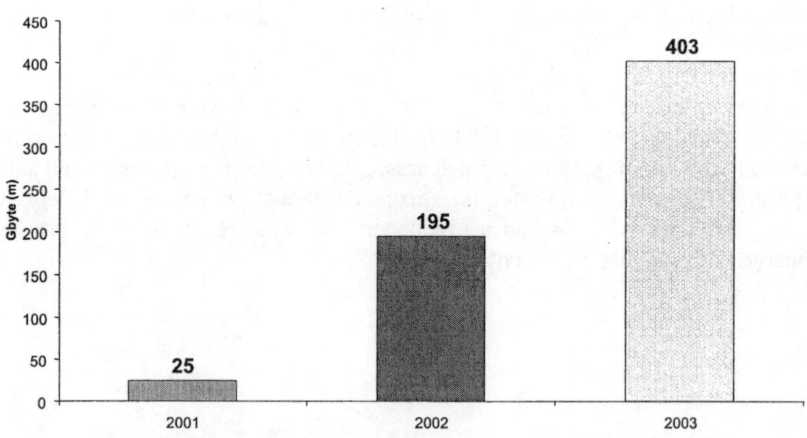

Fig. 4. Development DSL traffic

The data in the above diagram includes both the traffic volumes generated by the alternative network operators simultaneously offering Internet access services as ISPs, and the broadband volumes generated by the customers of ISPs without their own network.

1.6 Access to Deutsche Telekom AG's Subscriber Accesses

Apart from self-laid lines or radio accesses, for their subscriber accesses competitors use primarily DTAG's available subscriber accesses. At the end of 2003, 83 companies had signed contractual agreements with DTAG to this effect. More than 90 % of all competitor accesses at the end of 2003 were based on rented subscriber lines. In all, 1,349,848 such lines had been rented from DT at the end of 2003.

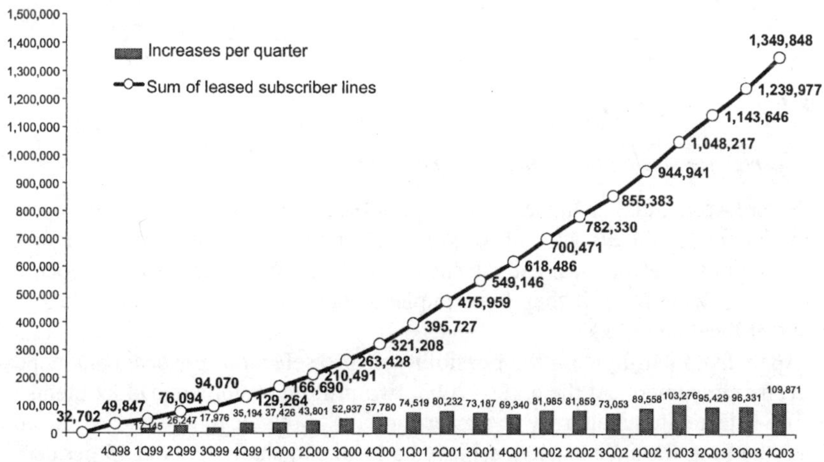

Source: Deutsche Telekom AG

Fig. 5. Development of DT leased subscriber lines

With an increase of 405,000, more subscriber lines were rented in 2003 than in the first three years of liberalisation in total. In view of the significance of narrowband and broadband Internet traffic in the fixed network the demand for subscriber lines on the part of the network operators is driven in the main by Internet use.

Fixed network traffic development

The narrowband traffic volume in the fixed network in 2003 was estimated at around 342 billion minutes and hence decreased compared with previous years. This is a result of the substantial traffic relocation from narrowband switched connections to broadband DSL connection. These high-speed accesses are usually characterised by particularly high usage rates. Also the increased use of mobile communications has an effect on the fixed network switched traffic.

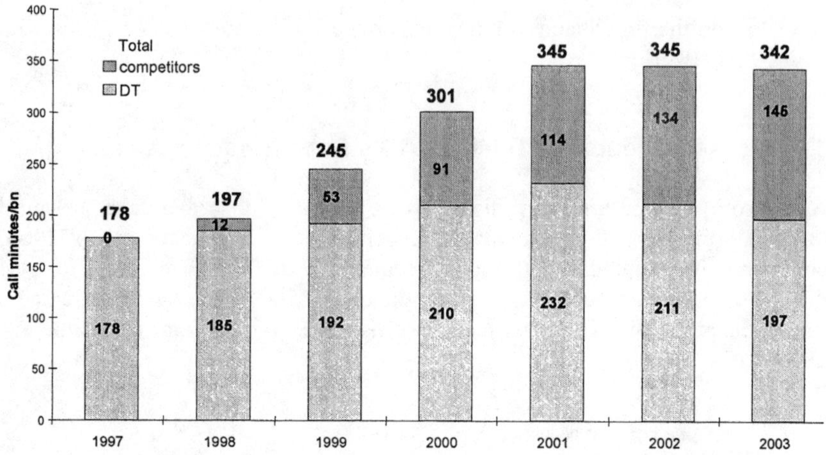

Fig. 6. Fixed line dial-up connection minutes 1997 - 2003

Carrier (pre)selection in the local network

Since April 2003 it has also been possible for consumers to freely choose a provider for local calls via call by call. Furthermore, since July 2003 preselection has been possible for local calls. Hence long-distance carriers can offer local calls on a nationwide basis if they have implemented all interconnection points in the relevant local networks.

Apart from call-by-call, the possibility of pre-selection for local calls appealed to many consumers. At the end of 2003, two million customers had let themselves be pre-selected to an alternative carrier for calls in the local network. At the end of 2003 the long-distance carriers achieved a local call share of about 15 per cent. To be added is a share of about 10 per cent of the local call minutes which were handled by the city and regional carriers. Consequently, the competitors handled roughly one quarter of all local calls by the end of 2003.

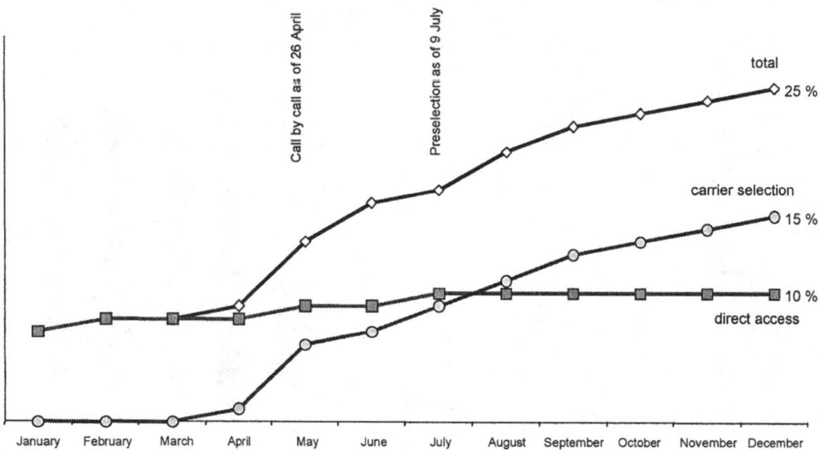

Fig. 7. Competitors' local traffic shares 2003

Market development mobile telephony service

At the end of 2003, 64.8 million customers could be reached in the German mobile phone networks (D1, D2, E1, E2). This corresponds to a penetration rate of 78.3 per cent and a high annual growth rate of nearly 5.7 million customers. The penetration rate in Germany is hence below the west European average of 83 per cent. However, in international comparisons it should be borne in mind that the German mobile radiocommunications statistics have been cleared from non-active prepaid clients. To what extent this also applies to other countries is difficult to assess.

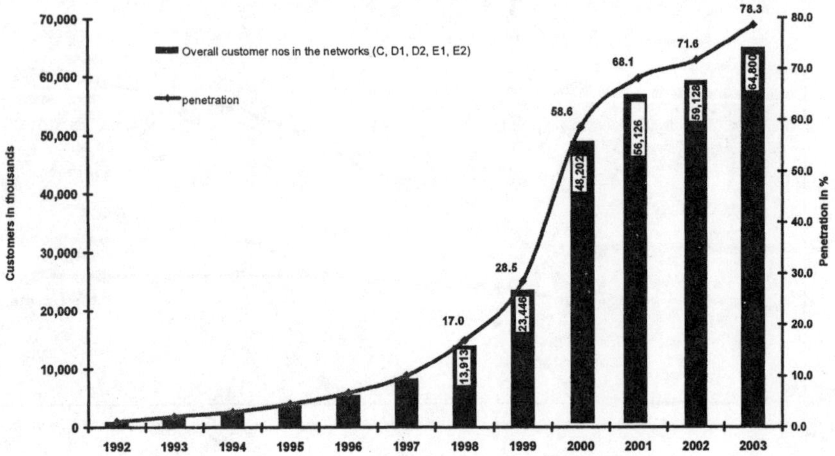

Fig. 8. Customer development (penetration and increases) in mobile telephone networks

The independent service providers' share of the mobile telephone service changed in the year 2003 to around 27 per cent of the overall customer base from about 28 per cent in 2002.

2 The New Telecommunications Act (TKG) and its Impact on Future Market Trends

2.1 Consultation and Consolidation Procedure

Market definition and market analysis are among the key issues concerning regulation of the telecommunications market in the future. Usually, significant market power needs to be determined before regulatory obligations can be imposed. Such obligations no longer apply quasi-automatically as under the old TKG but have to be imposed on a case-by-case basis as a result of a market analysis, whereby at least one obligation has to be imposed.

The relevant legal basis are sections 10 to 15 of the new TKG. Moreover, section 123 (cooperation with the Federal Cartel Office) and section 132(4) sentence 2 (determination by the President's Chamber) also have to be observed.

In contrast to the previous approach, two major changes have to be pointed to: on the one hand, the material point of departure is no longer section 19 of the Law Against Restraint of Competition but, based on Articles 14 – 16 of the Framework Directive, the Recommendation on relevant product and service markets and the Guidelines on market analysis and the assessment of significant market power which the utmost account has to be taken of. This means that market definition, in particular, has to be looked at from a different angle because there are 18 markets now that have been predefined whereas, in the past, the so-called demand-side substitution was to the fore. On the other hand, all future draft decisions will have to be submitted to the Commission of the European Communities who even has a veto right in this matter as laid down in Article 7 of the Framework Directive.

In coordination with the Federal Cartel Office and the EU Commission Reg TP has carried out comprehensive surveys on access to the subscriber line (market 11), interconnection services (markets 8, 9 and 10), fixed network telephony (markets 1 to 6), wholesale mobile services (markets 15, 16, 17), wholesale broadband access (market 12), leased lines (markets 7, 13 and 14) and broadcasting transmission services (market 18). Following evaluation of the replies concerning the aforementioned market segments there has to be a national consultation in each case; moreover, the necessary agreement with the Federal Cartel Office has to be reached. And an EU-wide consultation on the draft decision has to be held, as stipulated in Article 7 of the Framework Directive, whereby the EU Commission and the other national regulatory authorities have to be involved, before a determination is made, finally, by the President's Chamber.

2.2 Consistency Requirement

The consistency requirement is defined in section 27(2) sentence 1 of the Telecommunications Act. Under this provision, the Regulatory Authority has to take care that rates regulation measures are coordinated. The relevant procedures have to be applied for securing the consistency requirement (cf. section 132(4) of the Telecommunications Act). Hence consistency can be characterised differently. On the one hand, it is a criterion of rates regulation and a requirement concerning the quality of the contents of the rulings on prices respectively, on the other hand, it constitutes a coordination requirement under procedural law.

The question arises in what context the question of consistency has to be answered. In concrete terms, such question has to be answered by the responsible Ruling Chamber on the basis of a Ruling Chamber procedure, at the same time it is a mission of perspective for the Regulatory Authority to fulfil outside the Ruling Chamber procedure. This means that the Regulatory Authority's rules of procedure have to be amended and contents standards have to be developed and observed.

Three aspects, in particular, are of relevance concerning the details of such contents: firstly, the question of unjustified bundling has to be clarified; this is, in essence, the question of whether efficient competitors are able to emulate Deutsche Telekom AG's tariffs. Secondly, criteria for price squeeze analysis have to be established; in this context, the question has to be answered to what extent competitor-specific costs have to be taken into consideration. And thirdly, the significance of quantitatively and structurally appropriate wholesale tariffs has to be commented. In all these cases, it is important that the aspects of practicability and procedural economy respectively are duly taken into account.

3 Conclusions

Six years after liberalisation it is not only the companies that have to adapt themselves to changed market conditions. Regulation, too, has to respond to the increased complexity of the legal instruments provided by the new Telecommunications Act; at the same time the aspect of practicability must not come out of sight. And finally, all players in the telecommunications market must be given the necessary planning certainty.

It is all the more difficult to meet these requirements considering the fact that regulation will again and again have to overcome new problems caused by market forces driven by technological advances. In the year 2004, for instance, there was hardly any other topic that was more debated in the telecommunications market than Voice over IP (VoIP). IP telephony offers a chance for innovation and more competition. First commercial VoIP offerings to private customers have been available on the German market since year-end 2003. As early as April 2004 Reg TP started a consultation on VoIP in order to contribute to the creation of fair and reliable framework conditions fostering the spread of VoIP. Not only aspects concerning development of competition were raised in this context but also aspects of consumer protection and of public interest. Future discussions on VoIP will show, in particular, to what extent the existing regulatory framework will be sufficient to cope with the "new issues" in telecommunications or whether adaptations will have to be made for electronic communication services when this regulatory framework is reviewed again.

Innovation in the European and Japanese Mobile Telecommunications Industry

Richard Tee

1 Introduction 129
2 Mobile Data Services in Japan and Europe 130
2.1 Introduction 130
2.2 Mobile Data in Japan 130
2.3 Mobile Data in Europe 131
2.4 Explaining Diverging Developments 131
3 Examining the Innovation Networks in Europe and Japan 132
3.1 Introduction 132
3.2 The Strategic Networks Perspective 132
3.3 Sources of Differential Returns 132
3.4 The Strategic Network of Mobile Operators 133
3.5 Vertical Network Relations 133
3.6 Network Structure 134
3.7 Network Membership 134
3.8 Tie Modality 135
3.9 Conclusion 136
4 (Re)configuration of the Innovation System and Directions of Policy 136
4.1 The Japanese Mobile Innovation System 136
4.2 The European Mobile Innovation System 137
4.3 Are Reconfigurational Changes Desirable? 137
5 Overall Conclusions 138
References 138

1 Introduction

When we compare the development of mobile data services in Japan and Europe, we observe a number of important differences. These differences can be found along the lines of usage behavior - both quantitatively and qualitatively - as well as the role and structure of the various actors found in the overall market.

The chapter is structured as follows:

1. First, it will look at the development and diffusion of mobile data services in Japan and compare this to the situation in European countries. Also, it will very briefly summarize the various explanations forwarded to explain the differences that are found in both regions. The analysis takes market structural differences to be most relevant, in favor of explanations focusing on technical or cultural differences.

2. Subsequently, it will describe the respective systems from a strategic network perspective, looking at the central role of the network operator in Japan and how this contrasts with the situation in European markets. Using concepts such as network structure and membership, it concludes that the dominant position of the Japanese network operator has allowed the coordination of services to take place, whereas the European network attempts to stimulate innovation around protocols, comparatively unsuccesfully.
3. Finally, it will address the question how the systemic properties of the European innovation system, in terms of mobile data services, can be described and, if so desired, will also briefly take into account possible policy directions. Generally speaking, the Japanese network can be described as coordinated, operator-centric and service-based, whereas the European sytsem can be characterized as uncoordinated, vendor-centric and protocol-based.

2 Mobile Data Services in Japan and Europe

2.1 Introduction

During the last couple of years, European mobile network operators have been introducing mobile data services that offer information browsing as well as the ability to download games, ringtones, screensavers, etc. Examples of these services include Vodafone Live, i-mode, T-zones, and Orange World. These services can be interpreted as a response to the mobile services introduced earlier in Japan, of which the most cited example is i-mode, a service from incumbent operator NTT Docomo. Besides i-mode, there is the Ezweb/AU service from KDDI and also the Japanese version of Vodafone Live (formerly known as Jphone, offering JSky). This part will summarize the origins of mobile data in Japan, then looking at the situation in Europe. Finally, it will look at the various explanations that have been offered describing the differing paths of development in both regions.

2.2 Mobile Data in Japan

The case of i-mode is rapidly becoming a textbook example of successful innovation in mobile telephony. Introduced in early 1999, it currently enjoys a take up rate of about 90% (45 million) from all Docomo users (50 million)[1]. Besides i-mode, there is Ezweb/AU service from second operator AU, and the Japanese version of Vodafone Live (previously called Jsky).

[1] Figures for January 2006, see http://www.nttdocomo.com/files/presscenter/33_subscriber_0601.pdf.

Table 1. Subscriber numbers for the various services[2]

Customers by Carrier		Mobile Data Users		3G subscribers	
DoCoMo	50 658 500	i-mode	45 897 000	DoCoMo	22 015 000
KDDI/au	22 067 000	EZ-Web	20 105 400	KDDI	21 159 600
Vodafone	15 146 500	V-Live!	12 834 100	Vodafone	2 736 200
TU-KA	2 895 700				
Total	90 767 700	Total	78 836 500	Total	45 910 800

On the whole, these data services can be considered a success, given its visibility and usage rate among Japanese users. The initial 'killer-app' of i-mode were ringtones and screensavers. Later, success occured in the area of picture messaging (pioneered by Vodafone, Jphone at the time) and mobile games (mostly Java-based). Recent high profile development includes mobile payments (such as NTT Docomo's "Felica" system that can be used on public transport and for other types of contactless payment).

2.3 Mobile Data in Europe

Whereas the mobile services in Japan where instigated by the mobile network operators, the developments in Europe were originally propelled by handset and equipment manufacturers as well as Openwave, an American company specialized in "mini-browsers"[3]. Early developments in Europe centered on the WAP protocol, largely developed by vendors, in particular Ericsson. After the successes of SMS, it was imagined that WAP would extend this path by adding a protocol for both information browsing and other services. The reception of WAP and its subsequent development path contrasted sharply with the success of SMS. As a response to the unexpected slow take up of WAP, both from users and content providers, a number of network operators decided to implement mobile services in a comparable way to the Japanese operators. However, because of market structural differences, the actual implementation of these services has necessarily diverged in varying degrees.

2.4 Explaining Diverging Developments

A number of explanations have been offered describing the different development paths of mobile services in the two regions. Many of these explanations center around assumed differences in terms of cultural and social differences and technological distinctions. As forwarded in a number of other research papers, what is

---

[2] Figures for February 28 , 2006.
http://www.wirelesswatch.jp/

[3] For an insider's perspective on the deevelopment of WAP, see Sigurdson (2001).

lacking in analyses that focus solely on these factors is the inclusion of significant differences that exist in the area of industrial organization and overall market structure.[4]

3 Examining the Innovation Networks in Europe and Japan

3.1 Introduction

Building on the first part, this section will further examine the nature of the respective innovation systems, focusing on the industrial organization and overall market structure by employing a strategic network perspective. We will in particular look at the interaction between mobile network operator and handset/equipment manufacturers, which we believe constitutes one of the key components of the overall system.

3.2 The Strategic Networks Perspective

Among many other scholars, Gulati et al. have argued for the relevance of studying firms in the network of relationships in which they are embedded.[5] In the traditional neoclassical view, firms were regarded as atomistic actors competing for profits in an impersonal marketplace. In contrast to the neoclassical view, a relational approach is employed, where firms are conceptualized as operating in strategic networks which can differ widely and exist in a multitude of forms such as joint ventures, strategic alliances, and long-term buyer-supplier relationships.[6]

For companies, networks can potentially provide a number of advantages such as access to information and resources; learning and scope economies; and risk sharing and outsourcing. On the other hand, networks can be inhibiting as well, especially when a network is overembedded, meaning that firms in the network only relate to each other, thereby neglecting outside sources of potential utility.[7]

3.3 Sources of Differential Returns

Gulati et al. mention five sources of differential returns where network effects manifest themselves: industry structure, positioning within an industry, inimitable firm resources and capabilities, contracting and coordination costs, and dynamic network constraints and benefits. This part will focus on mainly industry structure

---

[4] Tee (2003).

[5] Dyer and Nobeoka (2000), Uzzi (1997), (Gulati et al (2000).

[6] See Ozman (2003) for an overview of different kinds network approaches.

[7] (Uzzi, 1997).

and its three relational characteristics: *network structure, network membership, and tie modality*. The first factor, network structure, relates to how network characteristics might influence the profitability of a firm. Examples of these characteristics include structural holes, network density, core versus peripheral firms, and structural equivalence. Second, network membership refers to the positioning of a specific industry in a broader context. Tie modality looks at the strength of connections that exist, looking within an industry itself but also across to suppliers and customer industries.

3.4 The Strategic Network of Mobile Operators

As a network standard, the European-created GSM is considered a great success. Yet data services are not nearly as successful as in Japan. This situation is potentially problematic for European operators, since their purchasing of the 3G licenses have been largely based on the idea that non-voice services will become increasingly important. Japan in contrast uses a network standard (PDC) that has not been adopted elsewhere. As mentioned earlier, however, the level of services is much higher and Japan has the highest number of mobile internet users both in terms of absolute numbers as well as percentages.

3.5 Vertical Network Relations

This part will look at the relationships between the operator and other actors involved in the overall system. It will mainly focus on vertical network relationships, which turn out to be of greatest relevance in explaining the differences between European and Japanese operators. Besides the network operator, there are two other organizations that can be part of the strategic network, handset and equipment manufacturers (also known as vendors) and content providers. The handset and equipment manufacturers supplies mobile phones to the operator and also produces the necessary equipment for the network to function; this is supplied in the form of equipment such as base stations. In most cases, these functions are performed by one company, though the division's output and success might of course vary. For instance, Nokia is Europe's largest handset manufacturer, but the biggest equipment manufacturer is Ericsson (Fransman, 2002). Though networks play a very important role in mobile communications, for the purpose of this paper we focus mainly on the handset supply issue. The content provider is relevant in the context of mobile data services such as WAP, Vodafone Live and i-mode. There are a wide range of content providers one can think of, for example news agencies or companies that provide games, images or ringtones.

3.6 Network Structure

The wide uptake of GSM has had great implications for handset and equipment manufacturers and in turn has brought with it repercussions for European network operators. Network operators are in principle tied to their location, though expansion of a network is of course possible via joint ventures and foreign investments. Vendors are much more mobile in the sense that they can lay out networks and distribute phones much more easily across countries and continents. The networking and vendor industry are both specialized and characterized by comparatively high entry barriers. All in all in Europe, the number of operators greatly outnumbers the number of vendors (approximately 50 operators compared to about 6 vendors). The situation in Japan is more or less reversed. Japan has only three operators compared, to a high number (more than 10) of handset manufacturers. These vendors are in a sense much more tied to their country since Japan is the only country that has taken up the PDC standard (though vendors are of course able to produce handsets for other networks).

We can argue that the structure of the network, exemplified by the ratio of operators compared to vendors, has had a great influence on the performance of the mobile network operator. In Japan, the ratio has favored the operator in terms of bargaining power, something that will become clear further on in the paper. In Europe, the opposite is true and the ratio of operators to vendors has favored the latter, which is also less tied to its region given the large uptake of the GSM standard. Next we will see if besides network structure other industry structure characteristics such as network membership have played a role.

3.7 Network Membership

In contrast to Japanese mobile operators, operators in Europe spend very few resources on research (Kano, 2000). Most European operators have outsourced technical operations to handset and equipment manufacturers. It is true also for Japan that operators let vendors take care of technical issues such as network roll out. On the other hand, Japanese operators still maintain large R&D labs that ensure the technical level of expertise is fully maintained. Consequently, operators in Japan are well aware of technical possibilities and have exploited these in different ways than operators in Europe. The comparative lack of technical resources is also exemplified in the way vendors function in the network.

Major vendors in Europe can work relatively independently from the mobile network operator. They decide themselves how many and which type of models they will release. Of course there is coordination relating to, for example, the roll out of networks vis-à-vis the type of handsets that are released. For example, handsets that supported EDGE were by and large not released until the first EDGE network upgrades were implemented. On the other hand, an Austrian operator proudly claimed to be the first country in Europe to have rolled out a 3G network at the end of 2002. However at that moment, no 3G phones were actually on the market. The fact that vendors are able to work outside of expectation or requirements of operators can be seen in the wide variety of phones that are on the market. This is the case both in terms of form factor as well as technical design. With regard to form factor, we see an assortisment ranging from the traditional "candy-bar" phones to the clamshell design and designs derived from those. Technical design refers to the type of protocols and services, and operating systems that are supported. Examples of protocols are SMS, EMS, MMS, WAP; services usually refer to a specific combination of protocols coupled with a set of technical features (such as the implementation of a color screen, camera etc), for example i-mode, Vodafone Live and T-zones. Recently we have witnessed radical experimentation of design by leading vendor Nokia which has expanded their range of phones with gaming and media devices, differing widely from phones that can be found in Japan. While the wide selection of designs might have been positive for vendors and consumers, it can be argued to have been detrimental for the uptake of data services. In Japan, by contrast, the variety of mobile phones is relatively small. The large majority of phones have the "clam shell" design (Meyer, 2002). Screen size has more or less been standardized. This is arguably a requirement for frequent usage of data services as well. This will be elaborated on further in the next section. All in all, we see how differences in network membership have also played a role in accounting for the differences in the performance of network operators in Europe and Japan.

3.8 Tie Modality

The keiretsu networks in Japan are a classic example of networks of firms working together closely in a variety of industries. We see a comparable type of development in the mobile phone industry. Centered around the largest operator, NTT Docomo, there is a circle of suppliers (vendors) that has been referred to as "Mova" maker. These vendors are guaranteed that their models will be used in the next lineup of phones. The rest of the vendors are referred to as "by-makers" and are not certain whether their models will be taken up (Meyer, 2002). The majority of phones sold in Japan are from Japanese companies. It has turned out to be difficult for non-Japanese firms to enter the Japanese mobile phone market. Even an established firm such as Sony has only recently changed status from "By" to "Mova" maker. While the same might have been true initially for Europe, non-European vendors have gradually gained wider market share, the most striking example being Samsung and LG.

The Japanese operator can be thought of as the central organization in the mobile telephony network. First of all, it chooses the handsets to be released from the suppliers. Second, it can also set the technical requirements to which the phones must comply, thereby facilitating the ability for operators to coordinate these with their data services. Third, the operator also has a degree of control over content providers. This holds most notably in terms of content as well as pricing. All Japanese operators have created a so-called micro-payment system which enables users to pay for services in a standardized, fairly secure way. Content providers who wish to make use of this micro-payment system must adhere to certain rules the operator has set. While the degree of control of the Japanese operator might seem unreasonably high, it has proven to be effective in generating a system for sustainable content creation, a necessity for the successful take up of data services many European operators are so desperately waiting for. On the other hand, the dominant position of NTT Docomo has received criticism, accusing the company of monopolistic practices and drawing comparisons to Microsoft in the sense that it is stifling innovation. It can be argued that in some ways the Japanese network can be thought of leaning towards becoming overembedded. Cooperation between European vendors and operators is much less embedded. As mentioned before, vendors work very much independently from the wishes of operators. Innovations done by the vendor do not necessarily correspond with the interests of the operator. As a result, none of the recent mobile internet initiatives that have emulated the Japanese approach have been taken up. With respect to tie modality, we can see how differences in strength and nature of ties have played a role in predicting differences in performance between operators.

3.9 Conclusion

In line with the claims of Gulati et al, we observe that a focus on industry structure provides us with explanations that can account for differences in performance between firms. We see this in several ways. First, when looking into industry structure we can witness how differences in ratio between vendors and operators has benefited the former in Europe and the latter in Japan. Second, the idea of network membership is exemplified by the differences in research as undertaken by operators, and the consequent results these have generated. Also, if we look at the "locus of innovation" in the respective strategic networks, we see differences; in Europe it is the vendor who is mostly responsible for this, while in Japan it is the mobile operator together with vendors. Finally, performance differences also become clear when we examine tie modality. First a discrepancy in the strength of the strategic network exists, which is high in Japan and comparatively weak in Europe. Second, great differences in the role of the operator can also be seen. The Japanese operator's role can be described as controlling and coordinating, in contrast to the operator in Europe that works in a much less coordinated environment.

4 (Re)configuration of the Innovation System and Directions of Policy

4.1 The Japanese Mobile Innovation System

As described earlier, the mobile network operator performs a central role in the Japanese mobile innovation system. We might characterize the Japanese (market) system as follows: The direction of innovation coordinated by one actor, being the mobile network operator. We might say the market is operator-centric in the sense that this actor has the highest leverage over other actors, in particular with respect to vendors. It is also most visible to the end users. In Japan, the service operator is the primary brand, rather than the handset maker.

A relatively closed system, Japan's market is characterized, for example, in the way particular handsets are tied – exclusively – to particular service providers. Overall, drawing a causal link between market structure and rate of innovation, we might say that the current system has benefited the overall system. The Japanese market has a very high level of technological advancement with high ARPU levels. At the same time, the dominant position of the operator has been criticized, especially the supposed monopolistic tendencies of NTT Docomo.

4.2 The European Mobile Innovation System

The European innovation system is characterized quite differently. Innovation is driven by consortia and individual actors, in particular handset and equipment manufacturers, where Nokia is dominant. Data services are largely based around open protocols like SMS, WAP and MMS. Besides SMS, none of the other protocols have generated much take up. Within this relatively open system, a number of operators have introduced Japanese walled garden types of i-mode-like services. None of these services, including for example Vodafone Live, i-mode Europe, T-Zones and Orange World, have managed to generate the levels of success enjoyed in Japan. This should not come as a surprise when we consider the way by which the strategic network is set up. Overall, the system might be characterized as vendor-centric, given its dominant role in the standardization processes surrounding protocols, and most importantly, its visibility to the end user as the primary brand.

Compared to Japan, the European market has lower level of technological advancement (on factors such as handset weight, size, display type, battery life, memory, features, etc.) and also has relatively low ARPU levels, in part due to the comparatively high levels of pre-paid users.

4.3 Are Reconfigurational Changes Desirable?

When we consider both paths of innovation, we observe marked differences in terms of the structure of the innovation system and the outcomes of both technology and services. These differences might lead us to ask if the current situation found in Europe needs to be changed, and if so, through what type of mechanisms. It is the position of this researcher that the advancement of mobile data services would fit well with the roll out of 3G networks, and more importantly are in line with the ambition formulated in the EC's Lisbon agenda.

It could be argued that the current lack of enthusiasm in this respect represents a classic case of a 'start-up' problem, where due to historical reasons certain trajectories were chosen in favour of others. An important question is whether the current market structures allow this start-up problem to be solved by itself, or whether outside measures are neccesary. Jeff Funk, well known for his research on mobile internet development in Japan, has made a number of suggestions in the context of the American market stimulating data services. Most of these could also be applied to the European system, which in terms of network composition is more similar to the American system than to the Japanese. Some of these suggestions include applying the notion of universal access and revenue sharing to mobile internet email and coordinating and stimulating push-based services, similar to the way the government has required location-based technologies for safety reasons.[8]

5 Overall Conclusions

This chapter has given a characterization of the Japanese and European innovation systems and which direct implications these systems have on technological outcomes. In the first section, we have looked at the development and diffusion of mobile data services in both regions. Marked differences have been observed in terms of development and subsequent reception of mobile data services. Briefly, we have looked at various explanations that attempt to clarify these different situations, which might focus on discrepancies in terms of culture, technology and/or market structure. This analysis has strongly focused on the last aspect, which is considered here to be most relevant. The second section examines the industrial organization of both markets and describes market structure according to a strategic networks perspective which we might refer to as innovation networks. The third section finally looks at systemic properties (i.e., components) of the innovation networks, thereby characterizing the respective innovation systems. Finally, it deals with the question as to what extent a reconfiguration of the respective systems is desirable. If the industrial European agenda is to accelerate the take up of mobile data services, systemic reconfiguration might be achieved by implementing a number of policy measures which might include broadening the concept of universal service and the coordination and stimulation of standards surrounding mobile email and push-based services.

References

DYER, J.H. and NOBEOKA, K. (2000) "Creating and Managing a High Performance Knowledge Sharing Network: The Toyota Case", Strategic Management Journal, 21: 345-67.

FRANSMAN, M. (2002), Telecoms in the Internet Age - From Boom to Bust to . . . ? Oxford: Oxford University Press

GULATI; NOHRIA and ZAHEER (2000), "Strategic Networks", Strategic Management Journal, 21: 203 - 215

[8] Funk (2004).

KANO, S. (2000), Technical Innovations, Standardization and Regional Comparison - A case study in mobile communications. The Telecommunications Policy Journal, Elsevier Publishers Vol 24, no. 4

MEYER, R. (2002), "Keitairetsu", Jan 2002 issue J@pan Inc
URL: http://www.japaninc.net/article.php?articleID=665, last accessed 29 Nov 2003

OZMAN, M. (2003), "Self-Organizing Inter-Firm Networks", working paper MERIT, http://www.merit.unimaas.nl/publications

SIGURDSON, J. (2001), WAP: OFF – Origin, Failure, Future, paper prepared for JETS, URL: http://www.telecomvisions.com/articles/pdf/wap-off.pdf, last accessed 29 Nov 2003

TEE, R. (2003). Contextualizing the Mobile Internet, Masters thesis University of Amsterdam, http://www.ecdc.info/publications/reports/2003_05_rt_mobile.pdf

UZZI, B. (1997), Social Structure and Competition in Interfirm Networks: The Paradox of Embeddedness, Adminstrative Science Quarterly, Vol 42 issue 1 (Mar., 1997), 35 – 67

Collective Dominance, Market Analysis and the 2002 EU Framework Directive: The Case of Mobile Access and Call Origination in Ireland

Chris Doyle[1]

1 Introduction ... 141
2 Market Definition ... 143
3 Market Analysis ... 144
4 Market Analysis: Single Dominance ... 145
5 Market Analysis: Joint Dominance ... 149
5.1 The Law and Economics of Joint (or Collective) Dominance ... 150
5.2 ComReg's Methodological Test for Joint Dominance ... 151
6 Applying the Test for Joint Dominance ... 153
6.1 The Degree of Market Concentration ... 153
6.2 The Incentive to Coordinate ... 153
6.3 The Ability to Coordinate ... 157
6.4 The Ability to Detect Cheating ... 161
6.5 Enforceability of Compliance ... 163
6.6 Actual and/or Potential Market Constraints ... 165
6.7 Pent up Demand ... 167
7 Conclusions on Market Analysis and Joint Dominance ... 167
References ... 168

1 Introduction

One of the first attempts to designate licensed telecommunications operators as having Significant Market Power (SMP) as a result of joint dominance under the 2002 EU electronic communications regulatory framework occurred in Ireland. The market involved was wholesale access and call origination on public mobile telephony networks, or market 15 in the European Commission's Recommendation (European Commission, 2003). The Irish National Regulatory Authority (NRA) the Commission for Communications Regulation (ComReg) sought to designate O_2 and Vodafone as jointly dominant in what it termed the wholesale MACO (Mobile Access and Call Origination) market.

[1] I should like to thank Peter Alexiades and Martin Cave for their input to this chapter. I am also grateful to officials at ComReg for advice and information, especially Leonie Allen, Caoimhe Donnelly, Isolde Goggin and Dermot Nolan. Finally I am very grateful to the editors of this volume. The author was an advisor to ComReg on the market review of mobile access and call origination in Ireland. All views expressed are his and do not necessarily reflect those of ComReg or other advisors to ComReg.

Following the adoption of the Framework Directive in Ireland, ComReg launched a six weeks consultation on 27 January 2004 to review the MACO market (ComReg, 2004a). Interested parties were invited to submit comments by 9 March 2004 on a number of questions pertaining to ComReg's preliminary findings. In this 'national' consultation ComReg suggested that there were competition problems in the relevant market, and in particular that the two largest mobile operators (O_2 and Vodafone) were jointly dominant.

"More importantly, however, there are strong indications that the aligned or consciously parallel behaviour identified by ComReg has resulted in a lack of effective competition at the retail level. For example, there appears to be little effective price competition between Vodafone and O_2 in the provision of their retail services. Further, despite the existence of the potential to secure a wholesale revenue stream from the provision of wholesale access and call origination services, particularly in the context of the spare 2G capacity of both undertakings, these operators have yet to conclude any indirect access, wholesale minutes/ capacity, MVNO or national roaming agreements with other network operators or service providers. ComReg is therefore of the preliminary view that the Irish mobile wholesale access and origination market and the retail mobile communications market is not effectively competitive." (ComReg, 2004a, paras. 4.118-4.119)

In response to the national consultation ComReg received submissions from eight interested parties, including strident opposition from the four licensed mobile network operators (MNOs) in the MACO market: 3, Meteor, O_2 and Vodafone.[2] The fixed line incumbent operator Eircom did not hold a public mobile telecommunications licence at the time of the national consultation and agreed with the preliminary findings of ComReg.[3] The other respondents were generally sympathetic to ComReg's case.

On 9 December 2004 ComReg published its response to the January consultation document and notified the European Commission of its decision, (ComReg, 2004b). In this document ComReg maintained its view that O_2 and Vodafone were jointly dominant in the MACO market. Throughout the consultation process ComReg had liaised closely with the Competition Authority in Ireland. ComReg 2004b also contained ComReg's notification to the European Commission of draft measures pursuant to Article 7(3) of the Framework Directive. Responses to the Notification could be made on or before 20 January 2005.

---

[2] At the time of the national consultation the operator '3' (a 3G operator owned by Hutchison) was not offering a national public mobile service.

[3] Eircom had previously sold its mobile operator Eircell to Vodafone. However, it was known to ComReg, and more widely, that Eircom was interested in re-entering the mobile market, possibly as a mobile virtual network operator (MVNO). At the time there were no MVNOs offering services in Ireland and this was a key fact in the review. The prospect of remedies triggered by a finding of joint-dominance that would include a requirement to allow MVNOs was therefore of interest to Eircom and others seeking entry into the retail Irish mobile market.

ComReg officials met with the Article 7 Task Force in Brussels to discuss the Notification on 10 January 2005. On 20 January 2005 the European Commission wrote to ComReg indicating that it did not have any serious doubts about the ComReg decision and proposed measure.[4] The mobile operators submitted further analysis and argument against the designation of SMP, and stated their intention to appeal the decision to the Electronic Communications Appeal Tribunal (ECAT) in Ireland (an administrative appeal body established in Ireland in accordance with the Irish Framework Regulations).

After several months of procedural debate, ECAT set the date for the appeal oral hearings on 13 December 2005. The oral hearings lasted only a little over one day, after which ComReg notified its intention to withdraw the controversial decision. The withdrawal was done on the grounds of procedural problems. Unfortunately the economic substance of the case was not discussed before ECAT.

In this chapter I present the analysis as applied to the structural characteristics of the market which formed the foundation for ComReg's case for joint dominance in the MACO market in Ireland. The paper is organised as follows. In section 2 market definition is discussed. Section 3 outlines the market analysis. Section 4 develops the analysis for single dominance and section 5 looks at joint dominance. Section 6 looks the test applied by ComReg for identifying joint dominance. Section 7 concludes with a discussion of the application of joint dominance under ex ante regulation.

2 Market Definition

In its Relevant Markets Recommendation the EU Commission recommends that NRAs should analyse the relevant market for "wholesale access and call origination on public mobile telephone networks" (European Commission, 2003). According to the Commission "regulation will not be warranted if there is effective competition on these markets" (European Commission, 2003, para. 20). ComReg's view in the consultation was that a relevant product market for mobile wholesale access and call origination could be identified and that the geographical market was the territory of Ireland. For the purpose of analysing the relevant market for mobile wholesale access and call origination ComReg identified and analysed the key elements of the market at the retail level, which is linked to the wholesale market.

ComReg's market definition was largely uncontested and the relevant product market was said to consist of:

1. all wholesale access and origination services provided by an MNO within the territory of Ireland;
2. all the MNOs; and

4 Correspondence SG-Greffe (2005) D/200269, signed by José Manuel Barroso, President of the European Commission.

3. self-supplied access and call origination services by vertically integrated MNOs.

According to ComReg there existed a services market for the retail provision of mobile communications services (a cluster of services was the term used by ComReg), which comprised access, basic voice services, including national, international and international roaming calls, supplementary and value-added voice services, and basic and value-added SMS. ComReg restricted its cluster of services to be voice and SMS–based services, and consistent with the principle of technology neutrality, these could be offered over 2G/2.5G and 3G networks. Pre-pay and post-pay mobile communications services were also argued as forming part of the same relevant services market.

The view that pre- and post-pay services were in the same relevant market was challenged by some of the MNOs in their responses to consultation. ComReg claimed that competition problems were more evident in the post-pay segment of the market. However, a number of respondents argued that if alleged competition problems were evident in the post-pay segment, then competitive constraints from the pre-pay segment ought to be relevant. For example, if there were a hypothetical monopolist of post-pay services it would face competitive constraints from firms offering pre-pay services, as these firms could relatively easily (that is at low cost) switch production by offering a post-pay service.

All parties in the consultation agreed that a hypothetical monopolist of post-pay services would be constrained competitively by firms offering pre-pay services. In arriving at this conclusion, the thought exercise considered firms that were comparable in terms of coverage – in other words, firms broadly similar in terms of vertical characteristics were (implicitly) assumed in ComReg's market definition assessment. However, as I show below, in its market analysis – which looked at actual firms in the market – firms offering largely pre-pay services were observed not to be sufficiently similar in a vertical sense. Therefore it was perfectly possible to have a market defined 'broadly' but for actors in the identified market to give rise to ineffective competitive constraints within sub-segments of that market.

ComReg stated that the geographical scope of the market for the provision of mobile communications services at the retail level, as well as for access and call origination services at the wholesale level was the national territory of Ireland.

3 Market Analysis

Having identified the relevant market relating to wholesale access and call origination on public mobile telephone networks in Ireland, ComReg conducted an analysis of that market to identify whether it was effectively competitive. In the EU regulatory framework this requires identifying whether any undertaking or undertakings is or are deemed to hold SMP in the relevant market.

The concept of SMP in the framework is equivalent to dominance in European competition law. However, a key difference between the framework applied to electronic communications networks and competition law is that the presence of dominance alone is sufficient to trigger obligations (remedies) in the new framework. Under competition law, for remedies to apply it is necessary for a competition authority to show that an abuse of a dominant position has occurred (in article 82 cases), or that a dominant position has been created or strengthened that would impede effective competition (European Community Merger Regulation, ECMR 139/2004).

The classic legal formulation for single dominance is found in United Brands v. Commission,[5] where the European Court of Justice held that a dominant position:

"... relates to a position of economic strength enjoyed by a undertaking which enables it to prevent effective competition being maintained on the relevant market by affording it the power to behave to an appreciable extent independently of its competitors, customers and ultimately of its consumers." (European Commission, 2002, paras. 72-74)

In situations where market shares are very high, they serve as a proxy for market power and as such there is a greater likelihood of finding SMP. The Court of Justice in Hoffman-La Roche v. Commission held that:

"... very large market shares are in themselves, and save in exceptional circumstances, evidence of the existence of a dominant position. An undertaking which has a very large market share and holds it for some time ... is by virtue of that share in a position of strength..."[6]

In competition cases, an undertaking is usually subject to a rebuttable presumption that it is in a position of single dominance if it holds in excess of a 50% market share.[7] Even a market share as low as 40% might, given the existence of other relevant factors, be supportive of a finding of single dominance.[8]

ComReg's market analysis concentrated on testing hypotheses about the presence of dominant undertakings in the relevant market. ComReg first considered whether any operator held a position of individual or single dominance. Following this, it assessed whether a group of operators were jointly dominant.

4 Market Analysis: Single Dominance

In the absence of any supply to third parties at the wholesale level, ComReg was not able to construct information about market shares for the relevant wholesale market; in the alternative therefore, it used information about the closely related retail market as a proxy of the relevant market shares of the parties. Absent access, the relationship between retail and wholesale mobile access and call origination is

---

[5] Case 27/76, [1978] ECR 207, para 65.

[6] Case 85/76, [1979] ECR 461, para. 41.

[7] AKZO v. Commission, Case C-62/86, [1991] ECRI-3359; approved in Hilti AG v. Commission, Case T-30/89, [1991] ECR II-1439.

[8] Op. cit., United Brands v. Commission; cf. Commission's Xth Report on Competition Policy (1980), at para. 50. See also SMP Guidelines, at paras. 75-76.

one-to-one, so retail and wholesale shares would be the same. The market share data examined by ComReg for the retail market are shown in Figures 1 and 2, using subscribers and revenues as the metric.

The share of market held by Vodafone, the largest operator in the retail market, was above 50% throughout the period of analysis. The market share of O_2, the second largest operator, was at or close to 40% over the period of analysis. The market share data appeared to be *prima facie* suggestive of individual dominance in the case of Vodafone, whose market share exceeds 50%.

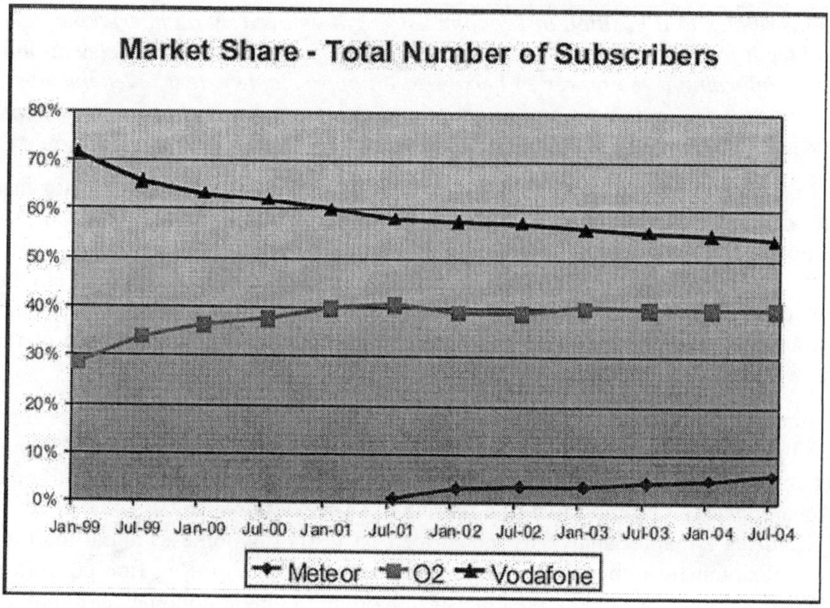

Source: ComReg, 2004b

Fig. 1. Market share – subscribers

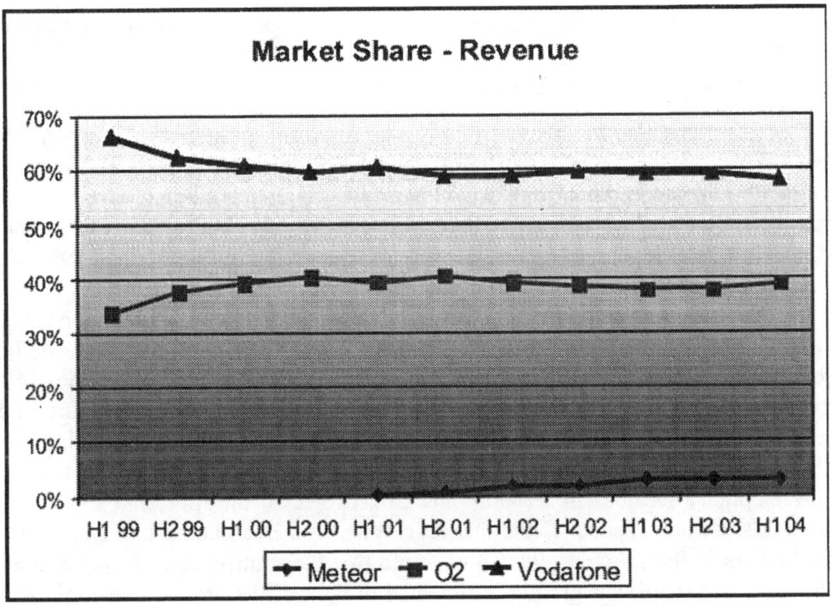

Source: ComReg, 2004b

Fig. 2. Market share - revenue

However, despite its relative importance, market share data cannot be relied upon, to the exclusion of other factors, as the basis for finding dominance. As the SMP Guidelines state:

"It is important to stress that the existence of a dominant position cannot be established on the sole basis of large market shares. As mentioned above, the existence of high market shares simply means that the operator concerned might be in a dominant position." (European Commission, 2002, para. 78)

To assess whether an operator holds a position equivalent to single dominance, ComReg needed to take into account a number of additional criteria, including:

- the existence of entry barriers and barriers to expansion (beyond strategic advantages which amounted to a 'first mover' advantage);
- economies of scale and/or scope;
- the size and distribution of competitors, relative to the largest firms;
- the existence of 'deep pockets' and access to capital;
- significant advantages in terms of advertising spend and other issues relevant to brand image;
- ability to leverage key aspects of overall size or economic strength in the relevant market;
- historical conduct suggestive of the existence of market power;

- economic performance relative to other market operators, taking into account efficiencies; and
- the extent of vertical integration, or the monopolisation of routes to market.[9,10]

ComReg was also made aware during its response to the national consultation that the European Commission had vetoed the Finnish NRA Ficora's finding of single dominance in the mobile access and call origination market, despite the fact that Telia Sonera had a market share in excess of 60% (European Commission, 2004). In this case market dynamics were examined, but unlike Ireland there were ten service providers operational in the market.

In assessing other criteria ComReg needed also to consider the 'Meridian Case',[11] which came before the Irish courts in 2000. In the Meridian case, the Irish courts were asked to take a position on the existence of single dominance in the Irish mobile sector. Despite Eircell (now Vodafone) having a market share in excess of 60% at that time, the High Court was not convinced by the available evidence that Eircell occupied a position of single dominance in the market for mobile telephony services in Ireland. The Court rejected the Meridian's contention that the structural aspects of the market resulted in Eircell being in a single dominant position. In particular, the Court found that the significance of the large market share of Eircell was greatly diminished in light of the dramatic decline of that share over a three to four year period, namely, a fall from 100% to 60% from the time when Esat Digifone (now O_2) entered the market in March 1997, and by the low barriers to expansion.

The Court, in its conclusion on dominance, stated:

"The reliance placed on the structural aspects of the market by the plaintiff does not appear justified on an analysis of this particular market. The significance of the large market share of Eircell is greatly diminished in the light of the dramatic decline of such share over such a relatively short period. The significance of the low number of competitors is diminished by the fact that Digifone is a strong company, well placed to exploit any laxity on the part of Eircell. I accept that the size of a competitor is not necessarily a relevant consideration in all cases for determining whether or not the firm can exert competitive restraint on a rival, but in this case, the strengths of Digifone are relevant in the assessment of Eircell's capacity to act to an appreciable extent independently of it. The significance of high barriers to entry in the market is vastly reduced by the fact that barriers to expansion are so low. Because of this, Eircell's capacity to act to an appreciable extent independently of its rivals is greatly reduced."

---

9 For discussions on the relevance of such factors, see for example: United Brands v. Commission, op. cit.; Continental Can, OJ [1972] L7/25; Hoffman-La Roche, op. cit.; Michelin v. Commission, Case 322/81, [1983] ECR 3461; see also Napier Brown-British Super, OJ [1988] L284/41. Refer also to the SMP Guidelines, at paras. 78-79.

10 The above criteria, to the extent that they are relevant, are also discussed in greater detail in the discussion on joint dominance below.

11 Meridian Communications and Cellular Three Limited v. Eircell Limited, High Court, IEHC 195, 5 April 2001.

In the Meridian Case the Court attached considerable significance to the observation that there were low barriers to expansion in the mobile market, which should constrain to an appreciable extent the ability of a firm with a high market share (in the Meridian Case, Eircell) from acting independently of its rivals. An important factor in the review was whether expansion costs were such that single dominance may be held by Vodafone and possibly also by O_2.

In the responses to the national consultation O_2 and Vodafone both stated that barriers to expansion continue to be low in Ireland. For example, it was stated that "even if entry barriers are said to exist in this market the effect of those barriers are mitigated by the absence of barriers to expansion",[12] and that "barriers to expansion are low".[13]

ComReg therefore formed the view that barriers to expansion were low and that it was not possible for Vodafone to act to an appreciable extent independently of O_2 and, similarly, O_2 was not able to act to an appreciable extent independently of Vodafone.

In the Meridian case the Court also relied to a large extent on other behavioural evidence, which suggested that Eircell was not then in a dominant position. Declining retail prices, evidence of 'leap-frogging' in tariff reductions, the large new and independent subscriber base, and the number and scale of innovations, were among the most important matters cited by Eircell as indicators of competition. This body of evidence, the Court concluded, supported the contention that Eircell's pricing behaviour was strongly constrained by competition from Esat Digifone.

At the time of the national consultation ComReg recognised that the absence of independent third party retail providers, apparent parallel pricing, and diminished growth in the market share of entrants all pointed to the possibility of joint (or collective) dominance.

5 Market Analysis: Joint Dominance

ComReg noted that a number of factors demonstrated that there was a prima facie case to investigate joint dominance. In summary these included evidence that, at the retail level:

- the market was highly concentrated;
- the market shares (in terms of subscribers) held by O_2 and Vodafone were relatively high (40% and 54% respectively at September 2004), and the difference between Vodafone's and O_2's market shares was less at the end of 2004 than it was in 2001;
- the market share held by O_2 was relatively stable at around 40% over the period 2001-04;

[12] Vodafone response to national consultation.

[13] O2 response to national consultation.

- the market shares of O_2 and Vodafone had been slowly approaching each other since the end of 2001, whereas before this date the difference in their market shares declined appreciably;
- new competition in the form of a third network operator (Meteor), which launched commercially during 2001, had not disrupted the market in terms of its key competitive characteristics, and at the end of 2004 remained on the fringe of substantial segments within the market;[14]
- evidence of a lack of competition was strongest in the post-pay segment of the market, where prices were regarded as relatively high and stable;
- there existed high barriers to entry;
- the profitability of O_2 and Vodafone was regarded as high; and
- there was evidence of parallel behaviour by O_2 and Vodafone in substantial segments within the market.

On balance, according to ComReg, the above factors warranted an investigation of whether Vodafone and O_2 occupied a position of joint dominance in the market.

5.1 The Law and Economics of Joint (or Collective) Dominance

Article 14(2) of the Framework Directive provides that an undertaking "may enjoy significant market power, that is, it may be in a dominant position, either individually or jointly with others." While it is clear that the European Court of Justice has recognised that "[a] market share of approximately 60% ... cannot of itself point conclusively to the evidence of a collective dominant position",[15] it is clear that combined market shares as high as 90% or more, as witnessed in the Irish market at the time of the national consultation, will normally be considered to "provide incentives for anticompetitive behaviour".[16]

A finding of joint dominance under the new regulatory framework should in principle follow the jurisprudence of the European Courts on article 82, as interpreted and implemented by the European Commission under its administrative practice. This is because the analysis under the new regulatory framework to be undertaken by ComReg should be based on a finding of collective dominance (i.e., an existing lack of effective competition), as occurs under article 82, which is expected to continue for the life of the review (the ex ante element). As is explained by the Commission in its SMP Guidelines, however, in applying the notion of collective dominance, NRAs "may also take into consideration decisions adopted under the Merger Control Regulation in the electronic communications sector".[17]

---

[14] The key segments in the retail market are pre-pay and post-pay. At the end of 2004 Meteor had a very low share of the post-pay segment.

[15] France & Ors v. Commission ('Kali und Salz'), Joined Cases C-68/94 and C-30/95.

[16] See, for example, Rhodia/Donal Chemie/Albright & Wilson, Case IV/M. 1517, at para. 61.

[17] Para 102, SMP Guidelines.

According to the case law, a finding of collective dominance may be based upon the existence of 'structural links' between the impugned collectively dominant firms, whose effect may facilitate the alignment of incentives among those coordinating firms. In the alternative, a finding of collective dominance can also be made in relation to an oligopolistic market whose structure is conducive to coordinated effects (i.e., anti-competitive behaviour) on the relevant market, without recourse to the existence of certain structural links.[18]

Although there was some discussion about structural links, ComReg placed little emphasis on these in the consultation. ComReg formed a view that it was structural characteristics in the market facilitating coordinated effects. The approach ComReg adopted was to apply the test for tacit collusion (joint dominance) that evolved in the Gencor and Airtours cases.[19] A key factor was the fact that the market was oligopolistic, as there were only four licensed operators. An oligopolistic market is more likely to have a structure which is conducive to coordinated effects, because firms are likely to become aware of their common interests and to anticipate one another's behaviour (Ivaldi et al., 2003).

5.2 ComReg's Methodological Test for Joint Dominance

In assessing the market and applying the test proposed for joint dominance as presented in Airtours, ComReg was to a considerable extent guided by Annex II of the Framework Directive, which provides for a test of joint dominance that is focussed on whether the structural characteristics of a relevant market encourage coordinated anti-competitive behaviour.[20] The criteria listed are expressed to be neither exhaustive nor cumulative, and said to be merely illustrative of the "sorts of evidence that could be used to support assertions concerning the existence of joint dominance".[21] The list of criteria is further repeated and elaborated upon in the SMP Guidelines.

The list reflects a broad range of issues that have been considered in the developing case law and in particular, the Court of First Instance (CFI) judgment in the Airtours case has been highly influential. The CFI set out the elements necessary to meet the legal standard for collective dominance.

ComReg approached its assessment of joint dominance in the market by taking into account EC jurisprudence and the SMP Guidelines. ComReg's investigation focused on whether the market structure is such that O_2 and Vodafone, in becoming aware of a range of common economic interests, could consider it possible,

---

[18] Para 94, SMP Guidelines.

[19] See Gencor v. EC Commission, Case T-102/96 [1999] ECR II – 753; cf. Airtours plc v. EC Commission, Case T-342/99, [2002] SCMLR 317.

[20] Recital 26 of the Framework Directive considers that a relevant market whose structure is conducive to coordinated effects means a relevant market that "encourages parallel or aligned anti-competitive behaviour on the market". Refer also to footnote 106 of the SMP Guidelines.

[21] Annex II, Framework Directive.

economically rational, and hence preferable, to adopt on a lasting basis a common policy on the market aimed at selling at above competitive prices, without having to enter into an agreement or resort to a concerted practice and without any actual or potential competitors, let alone customers or consumers, being able to react effectively.[22]

ComReg's analysis was based primarily on a review of the framework outlined below, which brought together the essential elements of economic analysis, EC jurisprudence and Commission administrative practice, as reflected in the SMP Guidelines:

1. *The degree of market concentration:* in particular whether a small number of undertakings accounted for a large share of the relevant market without any single undertaking being in an individual dominant position.
2. *The incentive to coordinate:* whether O_2 and Vodafone had and would continue to have an incentive to align conduct in the market in such a way as to elevate their joint profits by restricting production with a view to sustaining prices above those which would otherwise prevail.
3. *The ability to coordinate:* Whether O_2 and Vodafone had and would continue to have the ability to coordinate by aligning conduct so as to establish a consensus position (focal point). It was then argued that where there is an incentive for coordination and undertakings are able to establish a consensus position, the following conditions must be fulfilled for coordination to be sustainable over time,[23] namely:
4. *The ability to detect cheating:* the coordinating firms must have the means of *monitoring*, sufficiently quickly and precisely, whether the terms of coordination are being adhered to by other members of the alleged dominant oligopoly; and
5. *Enforceability of compliance:* there must be an *incentive* not to depart from the common policy on the market, of which the notion of retaliation in respect of conduct deviating from the common policy is an integral part.
6. *Actual and/or potential market constraints:* whether, in any event, the foreseeable actions of outsiders, such as actual and potential competitors, as well as customers, would not be able to jeopardise the results expected from the implicit coordination seen to be taking place.

An assessment of the existence of joint or collective dominance in the market took into account the overall impact of all these criteria, rather than being confined to the analysis of any single indicator of collective dominance.

ComReg applied the above test for joint dominance by relying largely, but not exclusively, on evidence available from the workings of the retail mobile communications market. This was because in the absence of any substantive transactions

[22] This language reflects that used by the CFI in paragraph 61 of its Airtours judgment, as well as by the Commission in its administrative practice.

[23] See, in particular, the Airtours judgment, Case T-342/99, Airtours plc v. Commission, ECR [2002] II-2585, and the SMP Guidelines.

at the wholesale level, the retail market would primarily provide the available competitive indicators to ComReg.

6 Applying the Test for Joint Dominance

6.1 The Degree of Market Concentration

In 2004 the retail mobile communications market was highly concentrated in Ireland with a small number of undertakings accounting for a disproportionately large share of the market. At the time of the market review there were four licensed vertically integrated mobile operators: '3', O_2, Meteor and Vodafone. Vodafone launched a digital GSM service in July 1993, O_2 launched a GSM service in March 1997, and Meteor launched a service in February 2001. '3', an operator having only IMT2000 (3G) spectrum) launched a testing service in September 2003 and only launched a commercial public service towards the very end of the review process. Throughout the review independent mobile service providers did not operate in Ireland.[24]

In September 2004 O_2 and Vodafone had a 94% share of subscribers in the retail market for mobile communications in September 2004, despite the presence of the third operator Meteor. ComReg observed that Meteor had been relatively unsuccessful, as its geographic coverage was less extensive than O_2 and Vodafone.

ComReg was required under the SMP Guidelines to adopt "a forward looking, structural evaluation of the relevant market, based on existing market conditions. NRAs should determine whether the market is prospectively competitive, and thus whether any lack of effective competition is durable."[25] A key event took place during the course of the review, when O_2 signed a national roaming agreement with Meteor, enabling Meteor to offer geographic coverage as good as O_2. The forward looking aspect of the review and the issue of entry and fringe competition are considered below in a separate section. However, ComReg concluded that despite this roaming agreement it did not expect competitive conditions to change appreciably in the near future and that there would remain a high degree of concentration in the market. This opinion was robustly challenged by all the mobile network operators.

6.2 The Incentive to Coordinate

A firm in an oligopolistic market has an incentive to align its conduct with other firms within an oligopoly if this would result in a profit that exceeds the level as-

[24] An independent mobile service provider is defined to be a firm that offers its own brand of mobile telephony service in the market, but is dependent on one or more of the mobile network operators for essential inputs.

[25] Para. 20, SMP Guidelines.

sociated with pursuing its own self-interest. The benefits of aligning conduct are enhanced in oligopolistic markets where there are entry barriers. In mobile markets there are considerable costs associated with entry.

However, while coordination offers a firm a higher profit at any moment in time, complying with a coordinated outcome means that a firm forgoes profit that would accrue through deviating from the conduct associated with the coordinated outcome. As competition recurs through time, the value to a firm of deviating is diminished as rivals would be expected to change their behaviour in response. Hence, the incentive to coordinate is largely based on assessing the merits of 'going it alone' versus 'aligning behaviour'.

It is the case that incentives for coordination are likely to be stronger when there are:

- few firms in an oligopoly, and
- key competitive indicators of the firms are relatively symmetric, (Ivaldi et al, 2003).

6.2.1 Few Firms in an Oligopoly

As the number of firms in an oligopoly sharing the collusive profit increases, it is clear that each firm gets a lower share of the "pie". A firm deviating would be able to obtain a large gain in profit by capturing a substantial share of the market. Consequently, there is a diminished incentive to coordinate as the number of firms increases in a market.

In the Irish market under review there were four firms, which in economic terms is generally regarded as few, rather than many (Selten, 1973). Furthermore, ComReg concluded that there was little prospect of entry before the next review (see section 6.6 below). More significantly, two of the firms, O_2 and Vodafone, commanded 94% of the market (in terms of subscribers) and ComReg took the view that the structure of the market was such that ComReg believed this was conducive to the alignment of behaviour involving these two firms.

6.2.2 Firms are Relatively Symmetric

It could be argued that O_2 and Vodafone were not symmetric, as O_2 had the lower share of the market at 40%, while Vodafone had 54%. Thus O_2 would face a greater incentive to deviate from any aligned behaviour aimed at enabling a collusive outcome. However, for deviation to be attractive the profit associated with a deviation needs to be relatively large or, if relatively small, durable. However, ComReg believed that Vodafone was in a position to react relatively quickly to a deviation by O_2 (deviations are discussed in more detail below). As O_2 had a relatively large share of the market, if it were to deviate it would sacrifice the excess profit from its existing customer base within a relatively short period of time due to the reaction of its larger rival Vodafone.

The proximity in market shares between members of an oligopoly can be a very helpful indicator of joint dominance. However, it is clear that relative symmetry of market shares is merely one corroborating factor in a finding of joint dominance.[26] The issue of the relative symmetry of market shares enjoyed by Vodafone and O_2 needs to be seen in its broader context. ComReg was of the view that, while the market shares of O_2 and Vodafone were not identical, the size of O_2's share of the market, especially given the overall size and structure of the market, was sufficiently large to indicate that it had an incentive to engage in behaviour that would and did give rise to coordinated effects. Moreover, when one is faced with a duopoly and the smaller of the two duopolists is already at such a market share level as to otherwise justify an enquiry into single dominance,[27] the concept of 'symmetry' should not be interpreted in a formalistic way which is equivalent to 'identical', as this would be unrealistic from a commercial standpoint. This is also borne out by the Commission's views on this particular issue, as set forth in a Position Paper submitted to the OECD in 1999:

"Symmetry of market shares can be an indication of similarity of incentives and of similarities in the scope for retaliation. However, market shares do not need to be completely symmetric in order for oligopolistic dominance to take place. It is quite conceivable that a merger will lead to one or more oligopolists being stronger than the other members in the oligopoly. In some situations there may even be a leader in the oligopoly. The important issue in the assessment of the symmetry of market shares is whether the market shares indicate a sufficient degree of similarity of incentives and retaliation possibilities. This can only be determined on a case by case basis."[28]

On balance, the proximity in market shares between Vodafone and O_2 relative to other competitors and in light of the maturity of the market, and especially given the existence of so many other factors facilitating coordinated effects, suggested to ComReg that the conditions conducive to the alignment of commercial behaviour were very much in evidence.

The difference in market shares between O_2 and Vodafone would be more relevant, for example, if it were shown to reflect more profound underlying asymmetries, such as in costs. ComReg noted that Vodafone and O_2 had similar portfolios of mobile network infrastructure and services in Ireland and had a similar population and geographic coverage. They also had the ability to source inputs and support services from the United Kingdom and other markets in their respective groups, resulting in economies of scale. In addition, these operators generally mirror each other's new products. Vodafone and O_2 had also been assigned an identical number of spectrum channels, namely, 36 channels of 200 KHz in the GSM 900 frequency band and 72 channels of 200 KHz in the GSM 1800 frequency band. All of the above would tend to suggest that their quality related costs were highly comparable.

---

[26] See Gencor, op. cit.

[27] See United Brands, op. cit.

[28] European Commission paper presented at OECD: "Oligopoly", Best Practice Round Table, October 1999, DAFFE/CLP [99] 25, p.218.

A number of other factors also had to be considered as playing a role in affecting the incentive to coordinate behaviour in the market.

6.2.3 Interactions

The incentive to coordinate is known to be easier where there is frequent interaction between firms in an oligopolistic market (Ivaldi et al, 2003). In mobile markets there is frequent interaction between the firms as there are many, often millions of consumers and many 'churning' customers on the demand side every month. ComReg noted that frequent interaction would be favourable for coordinated effects.

6.2.4 Growth in Overall Demand

The incentive to coordinate is also enhanced where the market demand for a product is expected to remain at current levels or grow. By contrast, if the market were expected to collapse, a firm aligning its behaviour would find it more appealing to deviate, as any reaction by its rivals in the future would be largely ineffective as the market would not exist. ComReg did not believe that the market demand for mobile telephony services would collapse or even decline significantly over the duration of the review.

6.2.5 Variability in Demand

A potential destabilizing factor for tacit collusion is demand variability. For example, if demand is known to be low next period but high this period, a firm may be tempted to deviate to benefit from current high demand knowing that subsequent punishment would be muted due to future low demand. ComReg argued that the macroeconomy would not lead to such variability in market demand over the period of the review. Furthermore, ComReg believed that total market demand for mobile telephony services in Ireland had not been subject to much unexplained variability over the period 1993-2003. The time path of demand, as measured by subscriber numbers, over the period 1993-2003, had steadily increased and followed a shape in form similar to a logistic distribution curve (Gruber and Verboven, 2001). Known seasonal factors tended to result in demand peaks and troughs within a year, with the final quarter in each calendar year (the Christmas sales period) notably generating the greatest additions in subscriber numbers.

6.2.6 Demand Elasticity

In relation to demand elasticity (Ivaldi et al, 2003) state "it has no impact on the sustainability of collusion. This comes from the fact that demand elasticity (and more generally, the shape of consumer demand) affects in the same way both the short-term gains from undercutting rivals and the long-term cost of foregoing future collusion." They go on: "Demand elasticity and buying power reduce the profitability of collusion; in addition, large buyers have more latitude to break collusion." Short term gains may be high (low) if demand elasticity is high (low), but also has a similar impact on the effect of the response (retaliation) on the part of the rival. Demand elasticity is therefore relevant for absolute profitability, but not for the factors related to the incentive to collude tacitly and the ability to sustain such collusion collusion.

6.2.7 Innovation

Another factor that could in theory undermine the incentives to coordinate is innovation. If innovation is significant in a market, then firms may compete by innovating which would enable them to gain larger shares of the market. ComReg accepted that there is a steady flow of new products and services into the Irish mobile market. However, these were observed to be innovations arising from industry wide developments, and were often the result of upstream manufacturers' efforts. ComReg did not see evidence of likely 'drastic' innovations[29] favouring one mobile operator which might undermine the incentive for coordination.

Having assessed these mainly demand side factors, ComReg concluded that there existed sufficient incentives for O_2 and Vodafone to coordinate behaviour in the market for wholesale mobile access and call origination in the Irish market.

6.3 The Ability to Coordinate

Even where there is an incentive to align behaviour so as to enjoy higher profits, it was important to assess whether O_2 and Vodafone had the ability to coordinate by aligning their conduct so as to establish a consensus position (focal point). By its very nature, the existence of a consensus position or focal point requires a high level of *transparency*. Where key information about rival firms, such as their conditions and prospects, is transparent this may facilitate reaching terms of coordination.

Furthermore, a consensus position is more likely to emerge where the focal point is *simple*. For example, if the rival firms sell a similar portfolio of products, retail prices for these products could comprise a focal point. Finally, *symmetry* assists in the establishment of a consensus position, as it enables rivals to identify variables that have in effect similar consequences.

ComReg believed that the focal point in the market had two dimensions and that these were closely related. Price was one dimension and the other was denying access to independent entities to upstream elements, such as wholesale airtime. Each is discussed in turn below.

---

[29] A drastic innovation occurs when a firm produces an innovation that leads to either a significant ('drastic') reduction in its costs, thus conferring it a permanent advantage over rivals in a market, or to a significant improvement in quality (or some other strategic variable).

6.3.1 Price as a Focal Point

ComReg believed that the first and most important dimension of the focal point was price. The ability to coordinate behaviour by focusing on price was likely to be enhanced if O_2 and Vodafone offer a similar portfolio of products. Where firms offer highly differentiated products it would make little sense for the firms to coordinate their behaviour, as monitoring deviations is too complicated (simplicity is sacrificed). ComReg believed that O_2 and Vodafone offered broadly the same portfolio of services in their retail clusters. This is due largely to the high degree of standardization in the GSM mobile industry. There is clearly brand differentiation, but the underlying generic products were identical and would therefore make it easier for O_2 and Vodafone to reach a consensus position on price, according to ComReg.

A key condition for tacit collusion to succeed is simplicity. However, ComReg acknowledged that there are many different tariff packages offered by O_2 and Vodafone in the market, and that the apparent complexity of these tariffs could work against the establishment of a consensus position. This point was raised by the mobile network operators in their responses to ComReg's position.

ComReg argued that where the underlying products are the same, it would be possible for firms to construct summary measures as a way of simplifying price information and thus enabling monitoring. Furthermore, ComReg was aware that at least one of the firms in the market did simplify price information to monitor price movements of its competitors. The most widely applied method to simplify tariffs in a mobile market involves constructing user categories (e.g. low user, medium user, etc.) and applying operator tariffs to these categories. For each user category it is possible to calculate the optimal pre-pay or post-pay tariff for each network operator (the optimal tariff being the one that offers the user category the lowest cost for his or her usage). The optimal tariff on a particular network for a user profile is then used to compute the 'minimum bill' over a period of time. Often minimum monthly bill figures are calculated, but sometimes minimum yearly bills are calculated (ComReg worked with both).[30]

ComReg analysed the price data over a period of 2001-04 and summarised the data in minimum monthly bills (MMBs). ComReg noted that the MMBs indicated relatively stable mobile pricing behaviour at the retail level by Vodafone and O_2 for high volume users and for those who benefit from being on post-pay tariffs. The relative stability of prices was argued to be both favourable for the establishment of a consensus position, and for the sustainability of such a position.

A necessary condition for price to be a focal point is transparency: in particular the tariffs of each operator need to be visible to one another. ComReg believed that the retail tariffs for voice telephony services and SMS text messages offered by O_2 and Vodafone to the vast majority of their customers were transparent, as they are published on the operators' websites. However, there are some customers (typically large corporates) that were able to negotiate directly with the operators,

[30] The appendix describes how minimum bills were constructed and used in the review.

and it could be argued, and was argued by the operators, that tariffs were not sufficiently transparent for price to form a focal point.

Although the prices established with a few large customers were determined on a bespoke basis, the ability for these customers to negotiate across networks made it highly likely according to ComReg that terms will be disclosed about operators' tariffs during negotiations. ComReg therefore believed that prices were transparent in the retail market. However, it did not present sufficiently robust evidence to uphold this position.

ComReg concluded that O_2 and Vodafone were able to focus on price because: (i) products offered are very similar; (ii) tariffs were transparent, and (iii) methods can be, and are, applied to simplify underlying tariff schemes.

6.3.2 Denying Access to Independent Entities to Upstream Elements

An inevitable consequence of focusing on the retail price to support coordination is the need to focus on closely related variables that could affect the retail price. ComReg argued that the other dimension of the firms' focal point involved the denial of access to upstream elements (on reasonable terms) that would enable independent entities to undercut O_2 and Vodafone. For example, this involved the denial of access of wholesale airtime to independent mobile service providers.

By denying access to independent service providers of wholesale airtime, this would effectively have the potential to lessen competition in the retail market (which amounts to vertical foreclosure). Denial of access may be constructive, through the setting of terms that make it commercially unattractive to an independent service provider to buy wholesale airtime.

ComReg argued that the denial of access to wholesale airtime is both simple and transparent, and therefore is an ideal dimension to the focal point. It was believed to be simple because the variable is binary: access is either granted or denied. It is transparent because each firm would be able to see whether the other firm had granted access.

ComReg observed that there was an absence of independent service providers in the retail mobile market in Ireland. Importantly, however, ComReg was aware of the existence of several entities seeking to offer an independent service provision. The fact that there were no independent mobile service providers, however, may be due not to denial of access by O_2 and Vodafone, but because O_2 and Vodafone were more efficient in retailing and that prices were at competitive levels. As a consequence the margins available to independent service providers would be insufficient to generate a return to cover costs – assuming access is made available on reasonable terms.

ComReg claimed that O_2 and Vodafone enjoyed high profitability in the Irish market, and therefore the terms offered to their own downstream affiliates were not being made available to independent service providers. ComReg also noted that independent service providers operated in a number of competitive mobile markets elsewhere in Europe (e.g. Finland and the United Kingdom), in many cases without the need for regulatory intervention. ComReg believed that it seemed improbable that such entities could not operate in the Irish market, even if it were competitive.

It could be argued that the granting of access to upstream elements to independent entities is much more likely to occur in a vertical market that is effectively competitive. If, for example, O_2 and Vodafone were not tacitly colluding, then there would be a strong incentive to offer access to an entity capable of delivering higher profits to the upstream access provider. ComReg believed that this incentive mechanism was a key factor behind the emergence of MVNOs (Mobile Virtual Network Operators) in the United Kingdom, where the national regulator authority Ofcom has not found any mobile network operator to have SMP in the mobile access and call origination market. The absence of MVNOs and other independent service providers in Ireland therefore suggested to ComReg a market that was not effectively competitive.

ComReg also noted that neither O_2 nor Vodafone had concluded a national roaming agreement with a new mobile network operator in the absence of a regulatory obligation to do so. During the review ComReg observed that O_2 signed a roaming agreement with Meteor (in the summer of 2004), which provided Meteor with national coverage. However, ComReg argued that this roaming agreement would not permit Meteor to affect materially its retail tariffs, because customers on the Meteor network would only occasionally roam into areas not covered by the Meteor network. ComReg was also of the view that the terms of this agreement were such that it did not favour Meteor in terms of price competition. Moreover, ComReg claimed, the agreement may not have been concluded in the absence of regulatory pressure exerted by ComReg during the market review.

ComReg did not present much substantial evidence to substantiate its claims about denial of access forming an important element of the focal point. Then operators in their responses attacked ComReg on this point, and suggested that the regulator had not shown convincingly that entrants were seeking reasonable offers from the networks. Indeed, according to the mobile network operators, airtime seekers and others were asking for extraordinary discounts and were using the review strategically to persuade the operators to grant these. The mobile network operators provided much detail about negotiations with access seekers, but this evidence did not persuade ComReg about the nature of the focal point.

While its claims about price as a focal point were quite strong, ComReg appeared to be in a weak position on the issue of access.

6.3.3 Countervailing Buyer Power

The ability to coordinate behaviour is also dependent on the position of customers in a market, and in particular whether a significant group of customers could exercise countervailing buyer power. Countervailing buyer power exists where, in particular, large customers have the ability within a reasonable period of time to resort to credible alternatives (e.g., not to purchase) if a supplier decides to increase prices or to deteriorate the conditions of supply.

ComReg took the view that there was insufficient countervailing buyer power in the market. The absence of independent service providers meant that countervailing buyer power could not be exercised at a wholesale level. At the retail level, as regards outbound mobile calls made in Ireland, there was no evidence available to ComReg to suggest that large (i.e. business) customers had been able to exert pressure on either Vodafone or O_2 with regard to prices. Furthermore, the MMB analysis in the appendix B showed that O_2 and Vodafone had consistently maintained higher prices than Meteor for high volume users.

If entities demanding access to mobile call origination and access services at the wholesale level were able to enter the market, it could establish some degree of countervailing buyer power. In the Finnish market, the European Commission had noted that

"the fact that SPs ask for competing offers from different MNOs, supports the view that SPs and MVNOs possess a significant degree of bargaining power vis-à-vis MNOs." (European Commission, 2004)

ComReg concluded that Vodafone and O_2 had the ability to co-ordinate their behaviour in the Irish marketplace, and that the focal point was multi-dimensional.

6.4 The Ability to Detect Cheating

For tacit collusion to be sustainable in an oligopolistic market it is necessary that firms can react quickly to changes in behaviour by other firms. The ability to respond quickly helps to reinforce the status quo by diminishing considerably benefits accruing through deviation. In the MACO market in Ireland ComReg considered two possible forms of deviation, each occurring along a specific dimension of the focal point: (i) price and (ii) access to upstream wholesale elements.

6.4.1 Price Deviations

A price reduction by firms in a tacit collusive situation is referred to as a deviation or cheating. A deviation would involve a reduction in MMBs, for example by lowering call charges or increasing bundled minutes or text messages in subscriber packages. ComReg claimed that movements in MMBs would be readily observable to all firms in the market and that deviations would be detected speedily. However, ComReg needed also to recognise that falls in MMBs could arise because of changes in tariff terms that were not the result of cheating. As a consequence, firms in the market would need to be able to distinguish cheating from non-cheating.

ComReg believed it was possible for O_2 and Vodafone to distinguish between price changes that represent deviations and price changes that are compatible with tacit collusion. ComReg claimed that there were two circumstances in the market that could give rise to lower MMBs compatible with tacit collusion.

1. Where cost movements across firms are highly positively correlated and the cost drivers are transparent. As the focal point changes as underlying costs change, then industry wide cost changes would result in changes to the focal

point prices. For example, if equipment costs in the industry were to fall, there would be a common incentive for each firm in the tacit agreement to lower tariffs. ComReg believed that some costs in the industry satisfied these criteria. However, it was acknowledged by ComReg that variability in correlated costs was unlikely to bring about swift changes in the focal point (i.e. prices) where firms tacitly collude. (Harrington and Chen, 2004).

2. Price changes that lower MMBs could also arise when firms' vary their tariffs to boost ARPUs rather than cut prices with the expressed intention of attracting subscribers from rival firms.[31] For example, ComReg argued that a reduction in tariffs by Vodafone across a number of user categories occurring in September/October 2001 arose as a result of Vodafone restructuring its Pay Monthly Options tariffs. While this may be perceived as a deviation along the price dimension, it is clear from the company's statements at the time that the restructuring was not designed to acquire an increase in customers. The company signalled very clearly to the market that the "New Pay Monthly Options are in line with the company's overall corporate strategy of focusing on customer retention rather than concentrating on programmes for customer acquisition".[32]

6.4.2 Deviation by Granting Access to Upstream Wholesale Elements

The other dimension of the focal point claimed by ComReg was the denial of access to upstream wholesale elements to independent entities. A deviation might therefore take the form of offering wholesale access to an independent service provider on terms that would enable the service provider to acquire significant numbers of customers. Such a deviation would be immediately visible – both from the appearance of a new service provider in the retail market, and by the fact that the new service provider would be offering prices to customers below the tacitly agreed position. Overall the market impact of granting access to independent service providers would be manifested in practice in its effect on retail competitiveness.

---

[31] It may seem odd to suggest that price changes which have the effect of lowering MMBs can boost ARPUs and hence, by implication, profitability. However, MMBs are constructed using profiles that have fixed quantity components. When a firm changes a tariff structure, there will be both substitution and income effects on demand that result in changes in the proportion of consumers across different user profiles. A tariff restructuring that boosts ARPUs effectively migrates customers onto higher MMB user profiles, by promoting increased demand for services via lower prices.

[32] Eircell Introduces a New Suite of Pay Monthly Plans Providing Significant Savings and Improved Customer Value http://www.vodafone.ie/aboutus/company/press/archive/release240901.jsp), 24 September 2001, as reported to the press on 4 October 2001, see Vodafone continues focus on higher value customers. (http://www.vodafone.ie/aboutus/company/press/archive/release041001.jsp)

6.5 Enforceability of Compliance

The European Court of First Instance has taken the position that the situation of collective dominance must be sustainable over time, namely, there must be some form of credible incentive not to depart from a common policy on the market. This means, for example, that each member of a dominant oligopoly must be aware that a highly competitive action on its part designed to increase its market share (a deviation) would provoke identical action by the others, so that it would derive no benefit from its initiative.[33] It is also an established economic principle that coordination between undertakings will be easier to uphold if deviations from a coordinated strategy – either explicit or otherwise – are more likely to be detected by other undertakings which have the means to punish deviation in an efficient and timely manner. This raises the issue of the enforceability of compliance, or alternatively the framework for deterring deviation or cheating.

In the CFI Airtours judgment it was pointed out that "the Commission must not necessarily prove that there is a specific 'retaliation mechanism' involving a degree of severity, but it must none the less establish that deterrents exist, which are such that it is not worth the while of any member of the dominant oligopoly to depart from the common course of conduct to the detriment of the other oligopolists".[34]

The issue of deterrence in the context of the mobile communications sector is best assessed by reference to whether the members of the collectively dominant position would revert to the normal conditions of competitive behaviour that would otherwise prevail in the absence of their parallel conduct.[35] In other words, at least at the retail level, the only deterrent or disciplinary mechanism required by Vodafone or O_2 to compel parallel behaviour would have been the implicit threat of their reversion to the normal conditions of competition.

"Indeed, a simple form of retaliation consists in the mere breakdown of collusion and the restoration of "normal" competition and profits. Firms then anticipate that collusive prices will be maintained as long as none of them deviates, but if one attempts to reap short-term profits by undercutting prices, they will be no

[33] "The notion of retaliation in respect of conduct deviating from the common policy is thus inherent in this condition. In this instance, the parties concur that, for a situation of collective dominance to be viable, there must be adequate deterrents to ensure that there is a long-term incentive in not departing from the common policy, which means that each member of the dominant oligopoly must be aware that highly competitive action on its part designed to increase its market share would provoke identical action by the others, so that it would derive no benefit from its initiative (see, to that effect, Gencor v Commission, paragraph 276)", para. 62, Airtours CFI Judgment (emphasis added).

[34] Para 195, Judgement of the Court of First Instance (Fifth Chamber, Extended Composition), 6 June 2002.

[35] In economics, in order for deterrence to be effective it needs to be credible. This means that, where a deviation occurs, the other firms in the (tacit) agreement would find it individually rational to apply sanctions (retaliate). Retaliation is understood here to mean a reversion to the conditions of normal competition.

more collusion in the future, at least for some time. Firms may then abide to the current collusive prices in order to keep the collusion going, in which case collusion is self-sustaining. This form of collusion has a simple interpretation: firms trust each other to maintain collusive prices; but if one of them deviates, trust vanishes and all firms start acting in their short-term interest." (Rey, 2002)[36]

6.5.1 Price

ComReg argued in its consultation that no deviation by O_2 or Vodafone had occurred along the price dimension since 2001. ComReg believed that a major factor deterring the use of price as an instrument of deviation was the threat of retaliation (or the common understanding that following a deviation normal competitive conditions would prevail). If one firm sought to acquire a significant increase in customers by lowering price, this would result in the other firm reacting with a similar response.

ComReg held the view that Vodafone and O_2 could respond to a price deviation by lowering retail prices, or by offering wholesale airtime on terms that would enable independent service providers to offer lower retail prices. One or other, or both, responses could be rational, if the other firm undertook a course of action which reduced the retail price of services provided over its network.

ComReg believed, however, that the most likely retaliatory response to a deviation along the price dimension, which would immediately be transparent, was via a reduction in price, as this could be implemented swiftly.

6.5.2 Access to Upstream Wholesale Elements

If O_2 or Vodafone were to deviate by offering access to wholesale elements that permitted an independent service provider to acquire significant numbers of customers in the retail market, it was claimed that retaliation would feature a reversion to normal competition. As stated above in the context of a price deviation, the non-deviating firm could respond by lowering its prices and/or by granting access to upstream wholesale elements.

Overall the potential reversion to the normal conditions of competition at the retail level was argued by ComReg to be a sufficiently serious disciplinary mechanism to sustain consciously parallel market behaviour.

[36] As has been borne out in the Commission's administrative practice, retaliation need not necessarily take place in the same product market as the deviation. See Commission Horizontal Mergers Guidelines, para. 55, where it is stated that: "If the coordinating firms have commercial interaction in other markets, these may offer various methods of retaliation".

6.6 Actual and/or Potential Market Constraints

Even a collectively dominant group of undertakings might be subject to significant competitive pressure, either from existing smaller 'fringe' competitors or potential new entrants.

6.6.1 Fringe Competitors - Meteor

The presence of a competitive fringe will, in theory, limit the scope for co-ordination, particularly where such competitors may be considered 'maverick' firms. The latter are generally reluctant to collude on account of, for example, their capacity reserves and different business strategies. It is important to note that while the presence of a competitive fringe may erode dominant firms' market power; it need not necessarily eliminate it completely.

ComReg believed that Meteor's past record in attracting subscribers, traffic and revenues demonstrated that it was unable to pose a serious competitive threat to Vodafone and O_2 and was unlikely to be able to adopt successfully the role of a "maverick" competitor. Indeed, despite posting lower prices than its bigger competitors, Meteor had not made any material impact on the pricing practices of Vodafone and O_2, particularly in the post-pay segment of the market. Meteor had only managed to gain a 1% share of the post-pay market since 2001, while it had an 8% share of the pre-pay segment of the market. The post-pay segment of the market in Ireland is more lucrative with both Vodafone and O_2 earning three times more revenue from their post-paid subscribers than from their pre-paid subscribers.[37] Although Meteor had a 6% share of the market in terms of subscribers it only accounted for 3% of the market in terms of revenue.

Vodafone and O_2 did lower their prices prior to the market entry of Meteor, there was little evidence to persuade ComReg that Meteor had made any significant impact on the retail tariffs of Vodafone and O_2 since its commercial launch, and especially since January 2001. The lack of full geographic coverage (until late 2004) was one factor that inhibited Meteor from gaining a greater share of the market and constrained its ability to compete effectively with Vodafone and O_2. This suggested that Meteor had not been a significant constraining influence on O_2 and Vodafone.

Geographic coverage differentiation appears to have been an important factor softening price competition, with services offered by the wider coverage networks being sold at higher prices than networks with smaller coverage. In that regard, Meteor's demographic and geographic coverage was appreciably less than that of Vodafone and O_2. Meteor's national roaming agreement with O_2 concluded in 2004 was expected by ComReg to help address this deficiency.

Meteor was also a 'late mover' into the market, having entered considerably later *vis a vis* third entrants in other EU Member States. Meteor entered after the

[37] Vodafone and O2 financial results to the year ended 31/03/04 reported ARPUs of €1,123 and €1,061 for their post-paid subscriber's compared to €361 and €357 for their pre-paid subscribers.

period of rapid expansion in the market. In particular, ComReg argued that the earlier entrants, Vodafone and O_2, with established subscriber bases and attractive on-net tariffs, had been able to sustain their advantage over the later entrant. In addition, the 'deep pockets' of Vodafone and O_2, particularly in terms of advertising and other forms of promotion, was claimed to constitute a significant competitive advantage over Meteor. That competitive advantage was reinforced by the fact that both Vodafone and O_2 benefited from having a significant amount of advertising available on Irish TV through the re-transmission of UK-origin TV broadcasts, significant purchasing power in obtaining handsets at attractive prices, pan-European branding, and the ability to shift costs offshore to their respective UK operations for certain types of "back-office" functions.

ComReg was presented with evidence in the response to consultation that Meteor's performance since 2003 had been improving and that it would have a market share of 10% of subscribers in 2005. However, ComReg asserted that even this market share would unlikely translate into effective market pressure in the absence of Meteor entering the lucrative post-pay segment, where it had only gained a 1% share in nearly four years of operation. ComReg believed that the national roaming agreement entered into between Meteor and O_2 might pave the way for potentially more effective competition in the market, but believed it would not provide a sufficient constraining effect on O_2 and Vodafone in the wholesale market.

ComReg's concluded that Meteor did not constitute, at least in the timeframe of the review, a significant competitive threat to the market positions of O_2 and Vodafone in Ireland. Even though Full Mobile Number Portability ("FMNP") was introduced in Ireland in July 2003, the evidence in the mobile market at the time of the review suggested to ComReg that the advantage enjoyed by Vodafone and O_2 would be of such significance that their current market positions in terms of their shares of the total subscriber base are unlikely to be eroded significantly by operators such as Meteor or '3' over the timeframe of the review.

6.6.2 Potential Competition and Barriers to entry/exit

The existence of entry barriers (legal restrictions, economies of scale and scope) affected the level of potential competition in the market. High entry barriers limit, in principle, the level of potential competition and may favour the adoption of collusive behaviour between existing competitors. This section examines potential competition and barriers to entry in terms of the impact of '3' on the market, potential 2G/3G network operators, and other entrants.

6.6.2.1 Impact of '3'

The 3G operator '3' entered the Irish market during the period of the market review and this made it difficult for ComReg to assess its likely impact. Nevertheless, ComReg formed the view that the arrival of '3' would unlikely have an effectively disruptive effect on the behaviour of Vodafone and O_2. As 3 possesses only 3G spectrum, ComReg believed this might actually strengthen the dominant position of Vodafone and O_2, as these operators would be able to migrate existing 2G customers onto 3G. On balance ComReg did not believe that '3' will be able to exert sufficient competitive restraints over O_2 and Vodafone.

6.6.2.2 Potential 2G/3G Network Operators

In addition to '3', ComReg was aware that a fifth network operator could enter the mobile sector by taking up the remaining unassigned 3G licence. However, ComReg believed that such a new entrant would face similar hurdles to '3' as outlined above. ComReg also considered the possibility that unassigned spectrum in the 900MHz and 1800MHz bands could potentially be made available for further expansion of 2G (GSM) services. A total of 2x26MHz of spectrum in the 1800MHz band and 2x13MHz of spectrum in the 900MHz band were unassigned at the time of the review. ComReg was of the view that entry barriers would deter potential 2G/3G network operators from effectively constraining Vodafone and O_2's joint dominance in this market.

6.6.2.3 Other Potential Entrants

ComReg considered whether other players may also enter the market via service-based competition, e.g. MVNOs or independent service providers, providing that appropriate wholesale services were available. ComReg did not expect such entry to occur during the review.

6.7 Pent up Demand

A controversial element of ComReg's case concerned what was termed 'pent-up' demand. Comreg took the view that the denial of wholesale access to service providers, such as MVNOs, was a feature of the tacit arrangement to sustain high prices at the retail level. Furthermore, ComReg claimed it had evidence of service providers seeking access to MNOs networks. However, in its consultation document it withheld from the public details about this pent-up demand. This met with robust criticism from O_2 and Vodafone, and during the appeal procedure ComReg was required to disclose to the affected parties its information on pent-up demand. ComReg claimed that its evidence demonstrated there was an unsatisfied demand for indirect access to the networks of O_2 and Vodafone.

7 Conclusions on Market Analysis and Joint Dominance

ComReg believed that the essential conditions warranting a finding of collective dominance, as set out in the judgments of the European Courts in Article 82 jurisprudence, as interpreted by the European Commission in the SMP Guidelines, and also in light of the criteria discussed by the Court of First Instance in Airtours, appeared to be satisfied in relation to the relevant market for wholesale mobile access and call origination.

ComReg applied a test for joint dominance using the three-pronged test established in Airtours.[38] The market had many of the characteristics that could give rise to coordinated effects. The combination of these factors made the Irish market particularly susceptible to collective dominance according to ComReg, and this was likely to continue in the absence of appropriate regulatory action. The implicit threat of a reversion to the normal conditions of competition at the retail level appeared to provide the type of retaliatory mechanism which could sustain the prevailing market conditions.

ComReg held the view that the Irish mobile wholesale access and origination market and the retail mobile communications market was not and would unlikely be effectively competitive without the application of appropriate regulatory measures. This view was robustly challenged by the affected parties, who submitted considerable opinion arguing against the ComReg decision.

The nature of evidence was such that it was not possible for ComReg to present evidence that would demonstrate without doubt that O_2 and Vodafone were tacitly colluding, as it did not have ready access to the detailed information such as costs. As in most cases involving alleged tacit collusion, the investigating authority needed to apply judgment and discretion.

The hurdle for passing the test for joint dominance in the new framework applying in telecommunications is relatively high, and as seen in the context of the Irish MACO case, it is very difficult for a regulatory authority to overcome this hurdle.

References

COMREG (2004a), Market Analysis – Wholesale Mobile Access and Call Origination, National Consultation 04/05, Dublin.

COMREG (2004b), Market Analysis – Wholesale Mobile Access and Call Origination, National Consultation (Response to Consultation Document 04/118a and 04/118b), Dublin.

EUROPEAN COMMISSION (2002), Commission guidelines on market analysis and the assessment of significant market power under the Community regulatory framework for electronic communications networks and services" 2002/C 165/03, 11 July 2002.

EUROPEAN COMMISSION (2003), Commission Recommendation of 11/02/2003 On Relevant Product and Service Markets within the electronic communications sector susceptible to ex ante regulation in accordance with Directive 2002/21/EC, C(2003)497, Commission of the European Communities, Brussels.

[38] Namely: (1) the existence of sufficient market transparency; (2) incentives for sustaining cooperation between the collectively dominant firms beyond the short term, which can be policed through an appropriate retaliatory mechanism; and (3) the inability of customers and existing/future competitors to affect the tacit coordination through their market behaviour.

EUROPEAN COMMISSION (2004) Commission Decision of 5 October 2004 pursuant to Article 7(4) of Directive 2002/21/EC (" Withdrawal of a notified draft measure") CASE FI/2004/0082: Access and call origination on public mobile telephone networks in Finland.

GRUBER, H. and F. VERBOVEN (2001), "The diffusion of mobile telecommunications services in the European Union" , European Economic Review, 45(3), pp. 577-88.

HARRINGTON, J. and J. CHEN (2004) "Cartel pricing dynamics with cost variability and endogenous buyer detection", September 2004, mimeo Department of Economics, John Hopkins University, available at http://www.econ.jhu.edu/People/Harrington/detectiondynamics9-04.pdf.

IVALDI, M., B. JULLIEN, P. REY, P. SEABRIGHT, and J. TIROLE (2003) "The economics of unilateral effects", March, Interim Report for DG Competition, European Commission.

REY, P. (2002), Collective dominance and the telecommunications industry, mimeo, 7 September, University of Toulouse.

SELTEN, R. (1973), "A simple model of imperfect competition where four are few and six are many", International Journal of Game Theory, vol. 2, pp. 141-201.

Credible Threats as an Instrument of Regulation for Network Industries

Justus Haucap, Ulrich Heimeshoff and André Uhde

1 Introduction 171
2 Objectives of and Reasons for the Regulation of Network Industries 173
3 The Light-Handed Regulatory Regime as Seen from the Perspective of the New Institutional Economic 177
3.1 Preliminaries 177
3.2 Light-Handed Regulation as an Administered Contract 177
4 Threat-based Regulation in Germany and in New Zealand 180
4.1 Electricity Regulation in Germany 180
4.1.1 Description of the Regulatory Regime and its Development 180
4.1.2 Evaluation 181
4.2 Telecommunications in New Zealand 183
4.2.1 Description of the Regulatory Regime and its Development 183
4.2.2 Evaluation 188
4.3 Electricity in New Zealand 189
4.3.1 Description of the Regulatory Regime and its Development 189
4.3.2 Evaluation 192
4.4 Overall Assessment 193
5 Summary and Conclusions 193
Appendix 196
References 198

1 Introduction

One of the major success stories for microeconomists involves the fundamental reforms that have taken place in network industries within the last 20 years. In many countries, the institutional framework under which utility and other network industries operate has changed significantly, involving aspects of privatisation, liberalisation and de- and re-regulation. However, even though the institutional framework has changed for the better in many countries so that efficiency has improved in most liberalised network industries, there is a still a vivid and ongoing debate about the proper balance between regulation and competition policy, or more generally, about the efficiency of various regulatory frameworks. Today, the focus of this discussion is on the questions of (a) how over-regulation can be avoided and (b) how and under what circumstances intrusive heavy-handed sector-specific regulation can be substituted by reliance on competition law or at least a more light-handed approach.

No matter whether more reliance is placed on ex post supervision or on ex ante regulation, one common policy objective is in any case to prevent the abuse of market power. It should be noted though that in liberalised network industries the focus of attention has shifted away from the prevention of behaviour that directly exploits consumers towards the prevention of exclusionary practices against (potential) competitors. Of course, exclusionary practices must ultimately also result in some form of exploitation in order to constitute a policy concern (see Vickers, 2005), but nevertheless the regulatory regime has to change if "exploitative abuse" (*Ausbeutungsmissbrauch*) is not directly but only indirectly prevented by enabling new entrants to compete for customers. If the latter approach is taken, the main concern are exclusionary practices or any "obstructive abuse" (*Behinderungsmissbrauch*) of market power. Hence, the focus of attention has also shifted towards the entry of new competitors and their survival, for which the ex ante regulation of some network elements is beneficial, as will be explained below.

At the same time, however, there is a severe risk of over-regulation, as is well known from the literature on the positive theory of regulation (see Stigler, 1971, Posner, 1974 and Peltzman, 1976). In general, there are two risks or types of errors associated with regulatory policy. A type-I error occurs if a market or firm is regulated intensely while it should be subject to a more light-handed approach (false positive), while a type-II error occurs if the regulatory framework leaves a firm too softly regulated while it should be regulated more intensely (false negative). Given the political economy of regulatory agencies, it is well accepted that there is a risk of over-regulation; put differently, regulatory agencies tend to place too much emphasis on reducing type-II errors, while they tend to underestimate the risk of type-I errors.

In order to avoid over-regulation, some countries have experimented with what might be called "threat-based regulation". The idea is very similar to the contestable market model, which predicts that *actual* competition may not be needed to discipline firms' behaviour, as long as incumbents are faced with the threat of market entry and *potential* competition (see Baumol, Panzar and Willig, 1981). As long as the threat of (hit-and-run) entry is credible, even a monopolist will offer its products at competitive prices. The concept of a threat-based regulatory regime is built on the related idea that actual regulation may not be necessary as long as the threat of regulation is credible. In that case, so the idea, even monopolies will abstain from setting monopoly prices in order to avoid being regulated (e.g., Glazer and McMillan, 1992; Bollard and Pickford, 1997; Acutt and Elliott, 2001a, 2001b). For the Government, such an approach may be attractive, because it avoids the costs of running a regulatory body. In addition, a threat-based approach to regulation may also reduce the risk of regulatory capture, which is usually associated with the establishment of sector-specific agencies (see Stigler, 1971; Peltzman, 1976).

Threat-based regulation may be regarded as one of many alternative institutional frameworks to govern network industries. Among the numerous possible institutional frameworks, threat-based regulation is situated between pure reliance on general competition law, which is usually applied ex post (with the notable exemption of merger control), at the one end and sector-specific ex ante regulation at the other end of the spectrum.

In fact, a threat-based or light-handed approach to regulation has been tried in both New Zealand and Germany. In New Zealand, quite a number of utility industries have been subject to what has been called the light-handed regulatory regime (LHRR). Until 1999, these industries included telecommunications, postal services, airports as well as gas and electricity (see Bollard and Pickford, 1997; Pickford, 1996). In Germany, a LHRR was applied to the gas and electricity industries until 2005 (see Brunekreeft, 2004), while the (fixed-line) telecommunications sector and other network industries have always been subject to some form of 'heavy-handed' regulation.

In both countries, however, the LHRR has now been abandoned in its pure form. In New Zealand as well as in Germany, a shift towards more traditional, heavy-handed approaches to regulation has taken place. However, while in Germany the LHRR has been replaced by a fully fledged heavy-handed regulatory framework, New Zealand has stopped halfway and maintained more elements of a light-handed approach.

The objective of this paper is to analyze the concept of a LHRR in more detail from a new institutional economics perspective. As we will argue in this paper, the success of any light-handed approach to regulation depends on the credibility of the threat of regulation, which is an essential part of the LHRR (see Bollard and Pickford, 1996, as well as Webb and Taylor, 1998). For this purpose, the rest of this paper is organised as follows: In section 2, we will briefly reconsider the objectives of and reasons for the regulation of network industries, before we relate the LHRR to the new institutional economic theories of regulation in section 3. Section 4 illustrates the concepts, successes and failures of the LHRR using examples from Germany and New Zealand, while section 5 summarizes our main findings and concludes the paper.

2 Objectives of and Reasons for the Regulation of Network Industries

In network industries, liberalisation is not the same as deregulation. While liberalisation means that institutional and other barriers to entry are removed so as to enable competitive entry, deregulation means that Government intervention into market processes is reduced (see Vogelsang, 2003). While in many industries deregulation and liberalisation go hand in hand, this is not true for network industries. Quite the contrary, some Government intervention is actually required to support market liberalisation and to facilitate competitive entry. More specifically, regulating access to so-called essential facilities or monopolistic bottlenecks helps to facilitate entry so that more competitive market structures can result in downstream or upstream markets that rely on the monopolistic bottleneck under consideration. While successful liberalisation processes also facilitate deregulation of some parts of the industry (namely, the potentially competitive markets), the terms and conditions for access to essential facilities need to be regulated. The very regulation of essential facilities helps in establishing actual competition in related markets that are potentially competitive.

If essential facilities are regulated in an efficient manner, market liberalisation should reduce the concerns about exploitative abuse of market power in potentially competitive upstream or downstream markets. At the same time, however, concerns about forms of obstructive abuse such as exclusionary practices will increase. Put differently, the deregulation of potentially competitive market segments requires the re-regulation of upstream bottlenecks.

At this point, however, one should keep in mind that competition is not an objective in itself. At least from a welfare economics perspective, it is only desirable if it leads to an increase in efficiency. Since the regulation of network access involves imposing obligations on network operators and owners, regulation always involves the redistribution of property rights. In market economies, however, any interference with or narrowing of private property rights needs to be justified on efficiency grounds, as the general presumption is that a system of private property rights maximizes welfare (e.g., von Weizsäcker, 1982 or quite fundamentally, North, 1990 as well as Barzel, 1989). Hence, any additional limitations on private property need to be justified before sector-specific regulations are imposed.

According to the *normative* neoclassical theory of regulation, such a justification is given as soon as there is some form of market failure due to some specific particularities of a given industry (see von Weizsäcker, 1982). Seen from this perspective, access regulation in network industries is already justified because some parts of network industries constitute resistant natural monopolies while other parts of the industry can be organised so as to facilitate competition (see Train, 1991). As is well known, a natural monopoly is given if the cost function is strictly subadditive in the relevant area of the market (see Panzar, 1989). In this case, one firm can produce a given quantity more cheaply than two or more firms. The classical case involves decreasing average costs due to economies of scale. In network industries, economies of density are also argued to give rise to natural monopolies (see Kruse, 2001). However, even according to neoclassical theory the pure existence of a natural monopoly in itself is not sufficient to warrant regulation. For regulation to be justified, a natural monopoly also has to be resistant to market entry. That means there must be barriers to entry which allow an inefficient situation to persist over time. This again is the case if market-specific investments (sunk costs) are necessary to enter an industry (see Kruse, 2001). The combination of subadditivity and irreversibility establishes a resistant natural monopoly, which is sufficient to warrant regulation according to the neoclassical theory of market failure.

Essential facilities are special cases of resistant natural monopolies. If access to a resistant natural monopoly element is necessary for potential competitors to enter a downstream or upstream market, this very element constitutes an essential facility. In other words, it has to be impossible to enter a market without having access to some specific facility for this element or asset to constitute a bottleneck (Knieps, 2001) or essential facility (see Areeda, 1990; Lipsky and Sidak, 1999; Rottenbiller, 2002).

Regulatory or competition policy problems result because the owner of such an essential facility will under a number of circumstances find it profitable to prevent or at least to slow down entry into the downstream markets by denying potential competitors access to these facilities altogether or by increasing access charges so as to render (large-scale) entry unprofitable. This fact warrants regulation from a neoclassical view.

From a new institutional economics perspective, however, the lack of contestability does not necessarily warrant regulation. Instead, the lack of contestability only provides a necessary, but not a sufficient condition for regulation. While contestable (or almost contestable) markets never warrant regulation on competition policy grounds,[1] the lack of contestability does not automatically imply that regulation is warranted.

From a comparative institutional economics point of view, regulating a resistant natural monopoly is only justified if Government intervention actually increases efficiency compared to all other available policy options (see Williamson, 1996; Dixit, 1996). Put differently, the net benefits of a given form of some necessarily imperfect regulation have to be larger than the net benefits of any alternative, including leaving the market unregulated. As long as it cannot be demonstrated with reasonable certainty that regulation increases efficiency, a market should be left unregulated. In addition, before a specific policy measure is implemented it should be compared to all other viable policy options. This comparative institutional approach and view of efficiency does not take the model of perfect competition as the relevant benchmark to guide economic policy, but rather operates with what Williamson (1996) has named the remediableness criterion.

While the remediableness standard poses a much higher hurdle on regulation to be adopted than the neoclassical market failure criterion, it has been largely accepted by now that not regulating resistant natural monopolies induces tremendous inefficiencies, also from a comparative institutional approach, whatever the relevant efficiency benchmark. First of all, a profit maximising monopolist will constrain output in order to increase prices, which results in an allocative efficiency loss, as some consumers will not be served even though their willingness to pay exceeds the marginal cost of serving them. Secondly, compared to a competitive situation some of the monopoly profit results from a transfer from consumers to producers (i.e., consumer surplus is turned into producer surplus). While such a transfer is not a problem from a pure efficiency perspective per se, this rent transfer can lead to further inefficiencies if it results in distributional conflicts in the form of rent seeking and other unproductive activities (see Tullock, 1967).

In order to avoid such inefficiencies and also for distributional and the corresponding political economic reasons, most jurisdictions have traditionally either regulated end-user prices (for example, the electricity sector in Germany), or the services have been directly produced and priced by state-owned enterprises (as in the telecommunications sector). In both cases, prices, service and quality levels, investments, and the like were largely determined by way of political reasoning.

---

1 There may be, of course, other forms of market failures such as externalities or information asymmetries, which may require regulation.

Usually, the objective of such a regulatory regime was the provision of certain services at low and often uniform prices on a nationwide level (see Knieps, 2003).

Quite generally, the regulatory reforms that have taken place in many of Europe's network industries were based on two insights. First, there is the insight that monopolies do not only lead to allocative inefficiencies, but due to the lack of competitive pressures they also induce so-called X-inefficiencies (see Leibenstein, 1966) or productive inefficiencies. A protected monopolist is likely to produce at inefficiently high cost levels. Second, it is now clear that not the entire supply chain should be viewed as a natural monopoly. A disaggregated view of the supply chain reveals that network industries can usually be disentangled into natural monopolies and potentially competitive segments (see Knieps, 1999). The latter parts can and should be liberalised so that competitive processes can lead to efficiency gains.

In summary, one is left to say that there is widespread agreement among economists today that leaving resistant natural monopolies unregulated is not efficient. Instead, some form of regulatory oversight is warranted. This also holds true from an institutional economics perspective. Hence, the thrilling question is not so much *whether* but *how* resistant natural monopolies should best be regulated. In designing an "optimal" regulatory regime, there are usually five broad objectives Governments look for when seeking to regulate network industries:

- Constraining companies from earning monopoly rents (allocative efficiency),
- maintaining incentives for network operators to seek productive efficiency,
- maintaining incentives for both operators and consumers to make optimal investment and consumption decisions (dynamic efficiency),
- minimising the administrative and compliance costs associated with the regulatory regime (transaction cost efficiency),
- minimising the extent to which the regime creates an incentive for lobbying, i.e., directly unproductive rent seeking (political efficiency), where transaction cost efficiency and political efficiency may be summarised as institutional efficiency.

The difficulty is that no one has ever designed a regulatory regime which is clearly superior on all of these fronts (and probably never will, as inherent trade-offs are involved). Hence, all regulatory regimes are necessarily imperfect. One of these necessarily imperfect regimes is the LHRR. Since most of the literature has focussed on the strengths and weaknesses of heavy-handed regulatory regimes, we will describe the features, strengths and weaknesses of the LHRR in order to facilitate a better comparison of the different approaches to regulation.

3 The Light-Handed Regulatory Regime as Seen from the Perspective of the New Institutional Economic

3.1 Preliminaries

While many countries have gone down a path of heavy-handed regulation of network industries (often even including the potentially competitive parts of the industry), New Zealand and Germany have experimented with what has become known as light-handed regulation. The LHRR-concept can be characterised as a specific regulatory regime in between the two poles of „reliance on general competition law“ at the one end and "sector-specific ex ante regulation" at the other end. As other regulatory regimes, the LHRR intends to constrain the abuse of market power which means that dominant firms should be prevented (a) from exploiting end users and (b) from discriminating against competitors. In general, the LHRR's objectives are comparable to those of other regimes including ex ante regulation and competition law approaches.

However, the LHRR diverges from both competition law and ex ante regulatory approaches in that the LHRR tries to include the best of two worlds. On the one hand, the LHRR tries to avoid the risk of regulatory capture and other costs associated with heavy-handed regulation by not establishing a sector-specific regulatory agency. On the other hand, the idea is to reduce type-II errors (i.e., lack of regulation when it would be welfare enhancing) by using regulatory threats to discipline firms for their actions. Hence, a LHRR relies on general competition law and enforcement through general cartel or competition authorities, complemented by a threat of more heavy-handed regulatory intervention if the firms under consideration do not "behave well". In order to monitor firms' behaviour, competition authorities under a LHRR are sometimes given greater investigatory power or the LHRR is augmented by sector-specific disclosure regulations or transparency requirements (as in New Zealand). Given that a LHRR involves more than pure reliance on competition law, the LHRR may be characterised as a regulatory regime somewhere in between strict sector-specific ex ante regulation at the one end and pure reliance on competition law at the other end (see similarly Williamson and Mumssen, 2000).

3.2 Light-Handed Regulation as an Administered Contract

In order to approach the question of how a regulatory regime in network industries can be evaluated in principle, one has to consider the objectives of regulation. From an economic perspective, market regulation should pursue an efficiency objective, whereas other social objectives (such as income distribution) should be pursued with other instruments (e.g., tax and social policy). In practice, however, this is generally not the case, for obvious political economic reasons (see, e.g., Baldwin and Cave, 1999), as universal service obligations in telecommunications demonstrate. Nevertheless, from an economic perspective, efficiency is the criterion against which regulatory interventions are judged.

In this context it is important to distinguish between static and dynamic efficiency. Static efficiency refers to a specific point in time and is given when social welfare is maximised at this point in time. In the simplest case (without transaction costs), static efficiency is given when all prices correspond to their corresponding marginal costs. In contrast, dynamic efficiency refers to a time period and requires that social welfare is maximised over the entire time period. Hence, a dynamic efficiency criterion also incorporates investment and innovation. Dynamic efficiency thus almost always requires a deviation from prices set at marginal cost.

The differences between static and dynamic efficiency are larger, the more important innovations and (other) specific investments (i.e., sunk costs) are and the higher the uncertainty associated with innovation and specific investments. Hence, there is a trade-off between static and dynamic efficiency when deciding upon the regulatory framework.

To achieve long-run market efficiency, the regulatory framework should on the one hand be designed to protect consumers from firms' potential market power (or its abuse), but on the other hand it should also protect firms from expropriation through the Government. From an institutional economics perspective, a regulatory framework can be seen as an implicit contract between operators and customers, which is administered by a neutral third party, the regulating authority or an independent court (see Goldberg, 1976; Crocker and Masten, 1996).

The more important specific investments are in an industry, the more important it is that the regulatory framework remain stable and predictable. An example may illustrate this point: Let us consider a project that requires a specific investment of X million Euros. The assets' expected time of use is t years, and both operational expenditures (OPEX) and capital costs (CAPEX) each amount to 1 million Euros per year. As the investment is specific and cannot be sold in the case of failure, the capital costs will in any case arise, independent of whether and how the assets will actually be used. The costs are therefore sunk. In contrast, the operational expenditures are avoidable by ceasing operation.

In this case, it is socially efficient to undertake the investment as long as there is a sufficient number of consumers whose willingness to pay is sufficiently high for the firm to cover its cost in the long-run. For this to happen, the firm has to be reasonably sure that the regulatory framework will not suddenly be changed and the firm will not be forced to lower its prices after the investment has taken place. From an economic perspective, this would be equivalent to expropriating the investors (see Sidak and Spulber, 1997).

Hence, for facilitating investment it is decisive to have a stable, predictable regulatory framework. The promise by the State not to "hold up" the investing firm ex post (i.e., after the investment has been made) and not to shift access rights by changing the regulatory framework must be credible. At the same time, however, the regulatory authority must possess a certain flexibility for considering new developments. Good regulation should therefore be open, transparent, consistent and accountable. This requires, amongst other things, open and transparent processes and criteria which can in principle be subject to judicial review (see also Levy and Spiller, 1994 as well as Williamson and Mumssen, 1999).

The ex post introduction of sector specific regulation can therefore be regarded as a Government breach of the implicit regulatory contract. Through this "hold up" by the State, firms' profits are socialised and redistributed even though the firms carried the original investment risk. Such a regulatory "hold up" (or the expectation of eventually being held up by the State) considerably reduces firms' investment incentives and, therefore, also harms consumers in the long run. In addition, it increases the overall governance costs as well.

Seen from this perspective, a LHHR may be regarded as a relational, administered, incomplete contract. While the essential facilities' owners implicitly promise to provide access to their infrastructure on fair and non-discriminatory terms, consumers (as represented through the Government) promise not to hold up investors by requiring price cuts once the firms have undertaken specific investments. Since it would be too costly, however, to contractually fix all possible contingencies ex ante (due to the contracting parties' bounded rationality), the contract will remain necessarily incomplete. The contracting parties will rather explicitly or implicitly agree upon contractual clauses or a "constitution of the contract" by which changes in future conditions lead to renegotiations. Since a third party (such as a sector-specific regulatory authority, a competition authority or an independent court) acts as an "arbitrator", one usually speaks of an „*administered contract*" (see Goldberg, 1976).

However, as is well known, due to bounded rationality and the lack of complete information (or other sorts of transaction costs), relational contracts cannot fully be enforced through third parties. Therefore, they need to be self-enforcing in order to successfully govern relationships (see Richter and Furubotn, 2003). Yet for a contract to be self-enforcing, there must be credible threats if either of the contracting parties deviates from the "spirit of the contract". That is, there must be a cost associated with deviation such as a punishment mechanism by the other party. The punishing party in turn must find it profitable ex post to punish the deviator (i.e., the threat to punish the other party in case of deviation must be credible).

In case of a LHRR, the self-enforceability of the relational contract may be ensured through a regulatory threat if (and only if) that threat is credible. Thus, in a first step the essential facility operators under consideration are required to offer competitors access to their network on a quasi "voluntary" basis (in the shadow of the threat of regulation). In a second step, the supervising authority uses its investigatory power and monitors (on behalf of consumers as well as actual and potential competitors) the terms and conditions according to which access to the essential facility elements is granted.

However, as network operators have incentives to overstate their true costs and to overcharge competitors, it is of vital importance that the threat of ex post intervention is credible. If the threat does not suffice to discipline network operators in their behaviour, the threat of regulation must be executed fast and swiftly. Problems arise as it is not easy for the Government to reliably verify all information regarding the circumstances at hand. Since external parties cannot be informed about all contingencies (see Williamson, 1985, p. 48), a LHRR will be more prone to type-II errors and too often leaves firms untouched when they should be subject to regulation.

4 Threat-based Regulation in Germany and in New Zealand

4.1 Electricity Regulation in Germany

4.1.1 Description of the Regulatory Regime and its Development

Germany's electricity markets have been completely liberalised since April 1998. In contrast to almost all other European countries, however, a sector-specific regulator was not established at that time. The so-called negotiated third-party access model was given preference over models with regulated access prices. That means, firms were supposed to enter into commercial negotiations and agreements about network access if one firm wanted to use another operator's network.[2] At the time of liberalisation there were almost 1000 distribution network operators on the low and middle voltage level and eight regional network monopolies on the high and highest voltage level (see Brunekreeft, 2003). Many network operators were also vertically integrated and active in either the generation or the retail segment, or in both (see Kumkar, 1997).

The duty to provide network access was enshrined in the new Energy Business Act of 1998 (*Energiewirtschaftsgesetz*). Additionally, the essential facility doctrine was also added to German competition law as §19 (s. 4) GWB (the Law Against Restraints of Competition). Hence, it was clear that network operators had to grant access to their networks unless there were objective reasons for denial (such as lack of capacity).

The access provisions were supposed to be enforced ex post by Germany's federal and state-level cartel offices, which had the task of monitoring firms' behaviour in gas and electricity markets and preventing dominant firms from abusing their position. Reliance was put on ex post control as is common in most other "regular" industries.

In addition, various industry associations, representing network operators, electricity generators as well as vertically-integrated companies and industrial users, were to negotiate general terms and conditions of network access in so-called association agreements (*Verbändevereinbarungen*). The first of these agreements concluded in May 1998 contained some almost obviously anti-competitive practices (such as distance-based access prices) so that the first agreement, called VV I, was replaced by a second one, VV II, in January 2000, which in turn was replaced by VV II+ in January 2002 (see Brunekreeft, 2003). However, even though these agreements facilitated some entry and (reasonably soft) competition so that prices initially fell, especially for industrial users, numerous disputes over network

[2] For examples of detailed analyses, see Bier (2002) and Brunekreeft (2003).

access emerged. In fact, the Federal Cartel Office took on quite a number of cases against various network operators for abuse of their dominance (see, e.g., Federal Cartel Office B11-38/01, Stadtwerke Mainz 2003 and B11-45/01, TEAG AG, 2003).

While electricity prices initially fell, they soon started to rise again and finally reached their pre-liberalisation levels again while electricity network operators achieved comparatively high profit margins. As some comparative studies have shown, network access charges remained especially high in Germany, both in absolute as well as in relative terms (see, e.g., VDN, 2005).

Between 1998 and 2003, the Government repeatedly declared that it was not satisfied with the (lack of) competition in the electricity markets and, on several occasions, threatened to replace the regime of ex post control by a more heavy-handed sector-specific approach based on ex ante regulation (see, e.g., Wetzel, 2003a, 2003b). While the industry always responded to the Government's threats, the steps taken were rather small, leading the Government to finally reach the decision that competition was not developing at a satisfactory pace and that the light-handed approach would be abandoned and replaced by ex ante regulation.

In addition, the EU Electricity Directive of June 2003 (*Beschleunigungsrichtlinie*) required Germany to establish an industry regulator and to replace the negotiated third party access regime with a regulated access price regime. Consequently, the new Energy Business Act which entered into force on 8 July 2005 has reallocated responsibilities from the Federal Cartel Office to the newly-created Federal Network Agency (*Bundesnetzagentur*) which regulates several network industries, namely telecommunications, postal services, electricity, gas, and rail. The methodology underlying the calculation of access prices must now be approved ex ante, while the detailed prices will be subject to ex post control (see Brunekreeft and Twelemann, 2005, for a detailed description and analysis).

4.1.2 Evaluation

Overall, there is widespread agreement among economists, legal scholars, and policy makers alike that the light-handed approach to regulation has largely failed in the German electricity industry (see Brunekreeft, 2003; Brunekreeft and Twelemann, 2005). As Brunekreeft (2003) convincingly argues in his analysis of threat-based regulation in Germany, one reason why the threat-based regulatory regime may have been doomed to fail may involve free-rider problems. The logic is as follows: If among 1000 network operators each operator knows that his behaviour is unlikely to have major impacts on the political choice of the overall regulatory regime, each operator has a stronger incentive to "misbehave" than in an industry with only one or a few regulated firms who know that the extent of regulation will very much depend on their own behaviour. If an industry consists of a single firm (as has been the case in telecommunications in many countries before deregulation), the firm's behaviour is also the industry's behaviour, so the firm may "behave well" in order to avoid being regulated. If, however, the industry consists of many firms, the industry's overall behaviour (and overall performance) is only marginally affected by a single firm's behaviour, and the individual firm is therefore less inclined to avoid industry regulation. While this analysis rests on the assumption that the same regulatory regime is applied to the entire industry without exceptions, this in fact reflects the political reality in most countries.[3]

In addition, two other important reasons for the regime's failure are (1) that the German competition authorities (the *Bundeskartellamt* and regional cartel offices) had severe informational disadvantages vis-à-vis electricity network operators and (2) the enforcement authorities were also lacking the necessary instruments to sanction inadequate behaviour even ex post. Therefore, the cartel offices had serious difficulties in confining monopoly pricing, as the cartel offices do not have the legal competence to prescribe detailed pricing rules or even to set prices. Of particular importance is that §19 (s. 4) GWB, which requires access to essential facilities under German competition law, does not contain a pricing rule (such as LRIC), but rather leaves pricing issues open. Nevertheless, the cartel office has the burden of proof to demonstrate that a given price level is excessive. What makes matters worse, the Cartel Office had to legally fight for access to the suspect firms' relevant cost data, and in a number of cases the Cartel Office did not obtain access to relevant cost data due to formal and procedural reasons so that the ex post supervision of price levels did not work well in practice (also see Schebstadt, 2004).

On the other hand, the threat of further (ex ante) regulation may have lacked credibility in Germany, given the responsible Minister's strong involvement in the electricity industry.[4] And finally, it is well known from game theory models dealing with trigger strategies that with incomplete information, threats have to be carried out from time to time in order to establish credibility (see Green and Porter, 1984; Abreu, Pearce and Stacchetti, 1986). However, once the threat of regulation has been exercised, the regime usually changes and there is little prospect for deregulation again, also due to well-known reasons of political economy. Hence, there is a ratchet-effect: Once a LHRR does not work satisfactorily, a move towards more heavy-handed regulation is difficult to reverse. Overall, the question remains how one may design a LHRR in which the threat of regulation is credible, but at the same time does not involve ratchet effects which shift the regime towards heavy-handed regulation. In this area, the New Zealand experience with utility reform provides some interesting insights, as we will demonstrate in the following section.

[3] Also see Glazer and McMillan (1992) for the general idea of regulation by threat.

[4] The then Minister of Economic Affairs, Werner Müller, held various positions with incumbent firms within the electricity industry before he was appointed Minister. After his departure from the Ministerial ranks, he once again took over the position of CEO at one of Germany's biggest energy firms.

4.2 Telecommunications in New Zealand

4.2.1 Description of the Regulatory Regime and its Development[5]

New Zealand is probably the country which has seen most economic reforms within the last 20 years, moving from a heavily regulated economy to one of the world's freest market economies (for a survey see Evans, Grimes, Teece and Wilkinson, 1996).

Among the many reforms that have taken place (concerning public finance, labour markets, social policy, monetary policy, trade policy and so on), the institutional framework governing network industries such as telecommunications has also changed dramatically. In fact, New Zealand experimented with what has become known as the light-handed regulatory approach to telecommunications, relying primarily on competition law and trying to avoid sector-specific regulation to the greatest degree possible. Interestingly enough, more than 10 years later the New Zealand Government abandoned its unique approach and moved towards a regime more closely resembling that of other countries. Most prominently, a Commissioner of Telecommunications has been established, who has the power to impose ex ante regulations. But let us start from the beginning.

In April 1987, *Telecom New Zealand* (*TCNZ*) was established as a so-called state-owned enterprise (SOE) under the *State Owned Enterprises Act 1986,* and the Government expected TCNZ to be run "as a successful business" and to be "as profitable and efficient as comparable businesses that are not owned by the Crown" (Galt, 1998). Seeking even more efficiency, the Government passed legislation in December 1987 which opened up all aspects of the industry to competition beginning on 1 April 1989. New Zealand thus became the first country to expose its entire telecommunications market to full, unlicensed competition since the early decades of the century, although the United States (US), the United Kingdom (UK) and Japan had each allowed partial competition by the 1980s. In August 1990, TCNZ was privatised.

New Zealand's LHRR was initially applied almost uniformly across all network industries and can be characterised by three main components:

1. First, reliance was placed on general competition law as given by New Zealand's *Commerce Act of 1986*. Under section 36 of the Act, dominant firms must not use their position to restrict entry to and/or competition in any market. Section 36 arguably applied to most of TCNZ's services and was argued to limit the terms that TCNZ can offer for interconnection. It has been enforceable through litigation, either privately (as in the Clear/Telecom dispute mentioned below) or at the initiative of the Commerce Commission, New Zealand's general antitrust authority. The maximum penalty a company can face under the Commerce Act of 1986 was NZ$5 million, but this has been increased to NZ$10 million in 2001.
2. Second, natural monopolistic elements of an industry are subject to specific "ring-fencing" accounting requirements and other information disclosure regu-

[5] The following part draws heavily on Galt (1998).

lations. Hence, section 36 of the Commerce Act 1986 was reinforced by the *Telecommunication (Disclosure) Regulations of 1990*, which required TCNZ to provide details of pricing for local access related services (including interconnection agreements), any substantial discounts, and financial accounts for local access-related activities. This was intended to provide incentives for compliance with the *Commerce Act* and to aid enforcement when required. These requirements aim at making the operations of dominant firms more transparent and therefore allow benchmarking to reveal eventual monopolistic pricing. These rules were intended to encourage self-regulation of the market (see Pickford, 1996, p. 211). The disclosure requirements were modified in 1999 so that TCNZ also had to provide separate financial statements and financial performance measures for its "local loop" and its "other telecommunications" businesses, including transfer prices between these businesses. Moreover, it requires the disclosure of any losses incurred in complying with the Kiwi Share obligations (to be described below), calculated on a forward-looking economic basis, and the way TCNZ recovers such losses.

3. Finally, under Part IV of the *Commerce Act of 1986,* the Government can on recommendation of the Commerce Commission take stronger action, such as introducing price control or other forms of regulation. Apart from the so-called Kiwi Share restrictions (see below), this threat was the only form of (indirect) price regulation imposed upon utilities. However, the Government subsequently showed extreme caution in reintroducing price controls, and until 2005 there have never been any goods subject to price control under Part IV of the Commerce Act.[6] In fact, the Government's threat to take further action was actually not regarded as being credible, so that TCNZ was only disciplined by competition from substitute goods (i.e., mobile telephony) and new technologies as well as by public pressures (see Bollard and Pickford, 1997, p. 497).

Underlying the concept of light-handed regulation was the Government's expectation that sufficient competition would develop so that an industry-specific regulator, as seen in almost every other country with telecommunications competition, could be avoided (see Galt, 1998). Residual competition disputes invoking the *Commerce Act of 1986* would then be resolved in the Courts. In the absence of sufficient competition, the price control provisions of Part IV of the Act could in theory still be invoked. Being aware of the pitfalls of the previous extensive use of direct price and other controls, all Governments from 1987 to 1999 saw them as a last resort.

At the same time, the 'Kiwi Share' held by the Minister of Finance represents New Zealand's approach to a universal service obligation, but also involves a

[6] The Commerce Commission recommended imposing price controls under part IV (section 52) of the Commerce Act 1986 on airfield activities of Auckland International Airport in August 2002. The Minister of Commerce did not follow that advice though, and no price controls were imposed. In July 2005, however, the Government followed the advice of a lengthy price control inquiry of gas pipelines and put two (out of eight) gas pipeline companies, Vector and Powerco, under price control under section 53 (2) of the Commerce Act (see Cabinet Office, 2005).

price cap on residential local telephone services (local access and calls). The Kiwi Share is essentially a contractual agreement between the Government and TCNZ that enables the Government to meet its social objectives. More exactly, the Kiwi Share enshrines a provision in TCNZ's constitution which requires that:

- increases in the price of residential access (including free local calls) do not exceed the 1990 price level (when TCNZ was privatised), corrected for the rate of inflation,
- the price for basic residential services in urban areas is also available in rural areas,[7]
- a flat rate (with unlimited free local calls) be available, and that
- the level of residential telephone service be as widespread as at the date of privatisation.

In December 2001, these so-called Kiwi Share obligations were updated to clarify that free local calls include standard calls to the Internet and fax calls and to slightly extend network coverage obligations and data capability requirements (14.4 kbps to 95% of existing lines and 9.6 kbps to 99% of existing lines).

A number of less formal regulatory measures have also always existed on top of the formal institutional framework. These informal measures included:

- statements of Government policy;
- the Government's active promotion of the New Zealand Telecommunications Numbering Advisory Group forum to seek agreement on telecommunications numbering issues between carriers, users and Government;
- publication of service quality indicators by TCNZ at the behest of the Minister of Consumer Affairs; and
- direct intervention by Ministers to urge a particular course of action.

The last of these measures, direct intervention, has always been an explicit possibility within the regime, and there have been two instances of explicit intervention. First, TCNZ was urged by the Minister at the time of privatisation to agree to reasonable terms of interconnection with Clear. Second, after the lengthy Clear/TCNZ dispute about local access terms had passed through the Courts without final agreement, the Government in 1995 told TCNZ and Clear that they would consider the introduction of price controls, and that the Baumol-Willig rule upon which TCNZ had insisted was not regarded as an acceptable basis for interconnection.

Following a lengthy Government review of light-handed regulation over the period from 1995 to 1996 (Ministry of Commerce/The Treasury, 1995), this opinion regarding the Baumol-Willig rule was formalised in a further Ministerial policy statement issued on 26 June 1996. The statement also expressed Government concern about slow progress with the introduction of telephone number portability be-

[7] The Government is not really enforcing this provision though, as, for example, basic services in Wellington (where TCNZ faces strong competition) are about 30% cheaper than in other (rural) areas where TCNZ does not face competition.

tween carriers, and referred to continued Government work to consider improvements to the regulatory regime while keeping the *status quo* for the meantime.

Certainly the single most striking characteristic of New Zealand's telecommunications industry has been the number of high profile competition disputes, which almost entirely relate to interconnection terms.

By far the most significant interconnection dispute was that of 1991-96 between Clear and TCNZ over local access terms. Important issues to be resolved included:

- TCNZ's initial insistence that its customers access Clear's customers through dialled access codes;
- TCNZ's refusal to provide direct dial in (DDI) lines to Clear to service its first local access customer, the Justice Department;
- TCNZ's insistence on Clear paying a levy for access to its network; and
- Clear's initial refusal to pay a contribution to TCNZ's network costs, including costs of meeting the Kiwi Share obligations.

Clear sought interconnection at incremental cost, with payments between the carriers being on a reciprocal basis. Part-way through the litigation, TCNZ offered pricing terms based upon the "Efficient Component Pricing Rule", often called the Baumol-Willig rule named after William Baumol and John Willig, who both gave expert opinions for TCNZ. This rule required Clear to pay TCNZ the opportunity cost of providing interconnection including any monopoly profit foregone by TCNZ from traffic lost to Clear. In fact, the case attracted extensive international comment (see Baumol and Sidak, 1994; Tye and Lapuerta, 1996; and Blanchard, 1994, 1995).

The Baumol-Willig approach was eventually held by the Privy Council (New Zealand's highest court based in London) *not* to breach Section 36 of the *Commerce Act*, after hearing arguments that it would lead to efficient price setting. While lower access prices would facilitate entry (and also reduce TCNZ's profits including the monopoly elements), entry may not necessarily be efficient in productive efficiency terms, as entrants might in theory have higher cost structures than TCNZ, thereby harming productive efficiency. The Court was apparently not persuaded that encouraging entry would ultimately have beneficial effects through the stimulus brought by competition to allocative and dynamic efficiency (see Pickford, 1996). The Court also mentioned the Government's ability to implement price control under Part IV of the Commerce Act if it were concerned that monopoly profits lay behind TCNZ's prices.

The Court held that Clear was entitled to damages for TCNZ's refusal to provide the direct dial service, but denied damages for TCNZ's initial stance on interconnection, prior to adopting the Baumol-Willig approach, even though that was also determined as breaching Section 36; Clear's refusal to contribute to TCNZ's network costs was itself found to be unreasonable.

The most important outcome of the Clear/TCNZ case was the upholding of the Baumol-Willig pricing approach to access. Subsequently, after a lengthy consideration of the issues raised (Ministry of Commerce/The Treasury, 1995), the Government declared that it was opposed to the use of Baumol-Willig pricing, because it had the potential to lessen competition (Luxton, Williamson and Kidd, 1996). The terms finally agreed upon by Clear and TCNZ set access prices at levels plainly below those implied by the Baumol-Willig rule (see Galt, 1998).

In total, it took six years after Clear entered the market before interconnection rates below monopoly levels were available to competitors. Even though market entry had taken place, prices had decreased and productive efficiency had improved right from the beginning of the reform process, these developments were tremendously accelerated after the major interconnection dispute had been settled. New Zealand had to wait for over six years for these additional benefits to arrive.

Given this time span, unhappiness over the light-handed regulatory regime increased within both the population and the Government. Even though the Government had generally preferred reliance on general competition law, the dissatisfaction with the light-handed regulatory regime manifested itself for the first time explicitly in a Government document. The coalition agreement, reached by the National and New Zealand First Parties in December 1996, stated the new Government's general direction for telecommunications policy as follows:

"*Amend Commerce Act to provide for penalties when actions are brought by parties other than the Commerce Commission. Our preference is for the Commerce Commission and the Ministry of Commerce to address competition issues. However, if this does not produce effective competition, Government must be prepared to ensure it does, firstly by producing Government Policy Guidelines on interconnect, transparency and number portability and if necessary by amending the Telecommunications Act.*"

No explicit action was taken, even though the junior coalition partner (New Zealand First) commissioned the Ministry of Commerce and the Treasury to report on possible improvements in the regulatory framework in 1998. However, no Government action followed that report, which recommended a full-fledged review of the interconnection regime.

It was only the next (Labour) Government which took action in March 2000 and set up a Ministerial Inquiry into Telecommunications. An independent task force was established which reported in September 2000. The Government responded in December 2000, and in December 2001 a new Telecommunications Act was passed.

The Telecommunications Act of 2001 now provides for a dispute resolution regime for access to regulated telecommunications services. In the first instance, the telecommunications industry is encouraged to reach commercially-negotiated agreements for access to telecommunications services. However, the Telecommunications Act of 2001 also provides that certain telecommunications services may be regulated if regulation will result, or is likely to result, in net economic benefits to New Zealand. The provider of regulated services will be required to provide that service to persons seeking access. If the parties are unable to reach agreement on the terms of that access, then the Commerce Commission (the general competition authority) is empowered to resolve disputes in accordance with specified access obligations.

More specifically, the Telecommunications Act of 2001 sets out a number of functions for the Commerce Commission to perform. In summary, these functions are:

- to make determinations on disputes over access to any of the designated and specified services listed in the Telecommunications Act of 2001;
- to make determinations on disputes over price between the access seeker and the access provider for designated services only;
- to conduct pricing reviews of its determinations for designated access services if requested to do so by a party to a particular determination;
- to undertake costing and monitoring activities relating to the Telecommunications Service Obligations (TSO) imposed on TCNZ as part of the Kiwi Share arrangement and to determine how these costs will be allocated to other industry players;
- to conduct investigations into the desirability of regulating additional services or amending the regulation of services where considered necessary, and to make recommendations to the Minister of Communications; and
- to decide whether to approve telecommunications access codes which have been submitted to the Commission by the Telecommunications Industry Forum.

In addition, a Telecommunications Commissioner has been established within the Commerce Commission (New Zealand's antitrust authority) under the Telecommunications Act of 2001. In March of 2002, the first Telecommunications Commissioner was appointed for a period of five years. The Telecommunications Commissioner has powers to ex ante regulate certain services. By now, the regulatory provisions of the Telecommunications Act of 2001 have been or are being used to regulate the supply of a number of telecommunications services. For example, the Commerce Commission:

- has completed determinations using the initial pricing principle for the supply of fixed network interconnection, retail wholesale services (excluding price-capped services), and residential wholesale services, and pricing review determinations which establish the price for these services using the final pricing principle;
- has completed the local loop unbundling review and following the Minister of Communications' acceptance of the Commission's recommendations, the introduction of Internet-grade bitstream unbundling and unbundled partial private circuit provision;
- has determined the net cost of the TSO for the financial years 2001/2002, 2002/2003, 2003/2004 and 2004/2005; and
- is currently investigating the case for regulating mobile termination.

4.2.2 Evaluation

In summary, the lengthy court dispute between Clear and TCNZ lead to growing discontent with the LHRR among consumers, competitors, and finally among politicians. To overcome the weaknesses of the regime, New Zealand has now installed a sector-specific regulator. However, in contrast to most other countries in Europe and North America, the regulator (the Telecommunications Commissioner) is situated within New Zealand's general antitrust authority, the Commerce Commission. Such an approach may reduce the risk of regulatory capture when compared to a sector-specific agency, but at the same time it helps avoid the inefficiencies of leaving bottleneck facilities de facto largely unregulated.

As it turned out, the dilemma with New Zealand's LHRR has been, at least partially, the same as in the German electricity industry. While the free-rider problem (analysed by Brunekreeft, 2003), did not play a role, as TCNZ was the only firm with significant market power in the industry, there was also significant political resistance in New Zealand to carry out the threat of price control. Consequently, the threat-based regulatory regime was not effective in constraining monopoly power, as the regulatory threat lacked credibility. In order to attain credibility, the threat finally had to be carried out. This, however, involved changing the entire regulatory regime. This is due to the fact that either TCNZ can be regulated or it can be left unregulated. If the regulatory threat is carried out, TCNZ faces regulation and the LHRR is de facto abandoned. Yet if the threat is not carried out, it may finally lose its disciplinary effect. This dilemma is difficult to overcome if it is only one single firm that is subject to this regime.

In contrast, if quite a number of firms are subject to threat-based regulation, the regulator can make an example of one or some firms in order to maintain the credibility of the threat, while not changing the entire regime and regulating all firms. At the same time, a regime which distinguishes between firms that behave well and firms that do not avoids the free-rider problem described above. This idea was further developed in New Zealand in the context of its reforms in the electricity sector and will be described in the following.

4.3 Electricity in New Zealand

4.3.1 Description of the Regulatory Regime and its Development[8]

Until 1987, electricity along with other utilities was a Government industry. While the central government owned all generation and transmission infrastructure, the local distribution networks were in the ownership of local government bodies or consumer trusts. Delivery of electricity was seen primarily as a public service, and there was no competition in any area of the industry. As mentioned above, the then Labour Government implemented a radical series of economic reforms in 1984, also involving corporatisation and privatisation of utility industries.

Since 1987, reforms of the New Zealand electricity industry have evolved through three main stages (a considerable degree of overlap between the stages): (1) Structural reforms by the Government, (2) industry-led reforms in the form of voluntary multi-lateral arrangements, and (3) specific areas of direct regulatory intervention.

---

[8] The following section draws heavily on Welson and Thomson (2002).

The first step in the reform process was the restructuring of the old Electricity Department into the *Electricity Corporation of New Zealand* (ECNZ) as a so-called state-owned enterprise (SOE) under the *State Owned Enterprises Act of 1986* to own and operate all transmission and generation assets. The SOE model, which was widely adopted in the public sector, requires state-owned companies to deliver outcomes comparable to private enterprise.

A wide range of further reforms were then implemented with the aim of creating a more competitive and efficient industry. The over-arching policy objective expressed through successive government policy statements was the need to achieve economic growth through efficient resource use, clearer price signals, and where possible with competitive markets. Accordingly, the Government's (both central and local) grip on the potentially contestable parts of the industry was reduced, and those parts were separated out from the natural distribution monopolies. The key changes included:

- the removal of statutory retail monopolies held by local electricity supply authorities,
- the separation of the transmission function from generation and the subsequent establishment of an SOE called *Transpower* as the national transmission provider which owned the national electricity grid in 1994,
- the separation of the Government-owned monopoly generator into two competing SOEs, *ECNZ* and *Contact Energy* in 1998,
- the mandatory separation of lines (transmission and distribution) and supply businesses as well as the prohibition of any entity operating in both sectors of the industry in 1998, and
- the sale of *Contact Energy* to private ownership and the further split of *ECNZ* into three competing state-owned generators (*Genesis Power*, *Meridian Energy* and *Mighty River Power*) in 1999.

Alongside these structural changes, a key feature of industry reform has been the emergence of voluntary multi-lateral industry-wide agreements including:

- the establishment of the New Zealand Electricity Market (NZEM) as a voluntary multilateral trading contract for the buying and selling of electricity on a spot market ex-post basis,
- the establishment of the Metering and Reconciliation Information Agreement (MARIA) to establish minimum standards for the measurement, recording and reconciliation of electricity distributed through transmission and distribution networks,
- the establishment of customer profiling and switching mechanisms as part of MARIA, to facilitate competition for retail consumers, and
- the development of the Multilateral Agreement on Common Quality Standards (MACQS) to enable grid users (rather than grid owner) to make decisions on common quality, including cost/quality trade-offs.

Have the reforms delivered efficiency and competition? To some degree they have achieved this aim. In particular, the establishment of NZEM resulted in the wholesale market becoming operational in 1996.

Other reforms were driven by overly simplistic or optimistic assumptions about the operation of the electricity industry, thus resulting in only partial success. For example, reforms targeted at developing retail competition (such as the split of lines and supply businesses) have been far more successful in delivering competition to major users than to private households.

In particular, the split of lines and supply companies did reduce the opportunity for lines companies to use their network monopoly to cross-subsidise their retail operations. In fact, vigorous retail competition emerged for large users. This, together with the introduction of profiling, appeared to deliver benefits for domestic consumers as well, with several retailers strongly competing for them (at least in urban centres) and offering price reductions and other incentives to switch.

However, the level of intense competition for retail consumers has been declining again. While a few small unaffiliated retailers emerged, these did not for the most part survive and most regions are still characterised by a single major retailer. Furthermore, instead of retailers being a counterbalance to generators' power, the market is now characterised by a high degree of vertical integration with all the major retailers being owned by generators.

Since the Government continued to be discontent with the development of competition in electricity markets, the Ministry of Commerce was asked to develop specific thresholds in April 1998 which would almost automatically trigger price regulation, but only for those firms who had breached these thresholds. The only exception from this automatic price regulation would be if the Commerce Commission itself regarded price regulation as undesirable even though the thresholds had been breached (see Ministry of Commerce, 1998).

As part of its electricity reform package, the Government also decided "in principle" to strengthen the threat of price control on electricity line businesses by empowering the Commerce Commission to apply price control within set criteria or thresholds.

The Government's fundamental objective was "to use the threat of price control to place enhanced pressure on electricity lines businesses so that they reduce their costs and set their prices based on these costs. The task is to do this in a way that mimics competitive pressure as far as possible and minimises regulatory costs. This means that the specific thresholds should be designed to:

1. gain some of the benefits of price control even when it is not introduced;
2. determine which line businesses should be subject to price control; and
3. reduce the regulatory uncertainty associated with the threat of price control" (Ministry of Commerce, 1998).

The Government also wanted the design and operation of the specific thresholds to be (a) credible, (b) consistent with the operation of actual price control, (c) specified in a way that mitigates the need for actual price control, (d) specified in a way that does not result in price control if it is not efficient and (e) unbiased so as not to penalise firms for factors outside their control. In order to enhance the credibility of the regulatory threat and to discipline firms in their behaviour, the consequences for a company of actual price control should be more onerous than the consequences for the company of the actions it would need to take to avoid breaching the specific threshold. In other words, the regime should not make it preferable for companies to be placed under price control.

The specific thresholds, which were designed as a mechanism for determining which if any firm should be placed under price control, were based on prices, rates of return and service quality. More specifically, each firm would be ranked every six months on three measures (from best to worst) and then scored (with "demerit points") on the basis of their ranking. The worse a firm's ranking, the more demerit points it would receive. The "best" firm, for each measure, would receive 0 demerit points while the worst would receive 100 demerit points. Then a weighted-average of the demerit points from each measure would be taken with the weightings reflecting the relative regulatory concern.

Hence, the minimum demerit points a firm could incur from the weighted average of the demerit points would be 0, if it were best on all three measures. Similarly, the maximum demerit points a line business could incur from the weighted average of the demerit points would be 100, if it were worst on all three measures.

If, on average, a firm's total demerit points for price, rate of return and service quality were equal to or exceed 75 over a specified period, its network access prices would be placed under price control. The initial measurement period was thought to be two years. That is, if a firm's average demerit points equalled or exceeded 75 over four consecutive six-months periods it would be subject to price control. The measurement period would be extended every six months until it reached five years, under which case it would be maintained as a rolling five-year average. To avoid price control, a line business that incurs 75 demerit points in any given six month period would therefore need to incur less than 75 in the future. An example, as developed by New Zealand's Ministry of Commerce, is given in the Appendix.

4.3.2 Evaluation

The idea of the threshold regime described above was to simulate the pressures that companies would face in competitive markets. Firms in competitive markets which cut costs by improving efficiency are able to earn a higher rate of return if they improve their efficiency faster than their competitors. Companies in competitive markets that cut costs by reducing the quality of their service will receive a lower price for their product. Moreover, in a competitive market, companies find that it is necessary to continually increase their efficiency and/or improve their service quality. Failure to do so means that they will progressively find it harder to maintain - let alone increase - their rate of return (as they will find that their competitors are able to undercut their prices).

While this innovative system has been discussed intensively in New Zealand, it has not yet been implemented. Instead, there is still an ongoing debate about the appropriate thresholds and mechanisms (see Commerce Commission, 2003, 2004).

Unfortunately, this approach has not been tested in practice yet. Form a theoretical perspective, the approach appears promising though because it would avoid a free-rider problem without regulating all firms in a heavy-handed fashion. In addition, the threat of regulation is credible, as it is exercised quasi-automatically once certain thresholds are breached.

Instead the Government installed a sector-specific regulatory agency in 2003, the Electricity Commission. The Electricity Commission has the task of guaranteeing an efficient and fair supply of electricity (whatever fair means in this context) for all consumers. While the Commission ought to provide incentives for efficient investment behaviour, it may not set prices. The power to impose price controls still rests with the Commerce Commission, New Zealand's general antitrust authority. At the same time, the exact thresholds for price control of electricity lines are still being discussed (see Commerce Commission, 2003, 2004). This means that the Electricity Commission lacks the powers other regulators usually have, making it a rather toothless beast. Since it is also not yet clear as to how the Commerce Commission will use its powers, it remains to be seen in which direction this hybrid regulatory regime will develop.

4.4 Overall Assessment

As everything else, threat-based regulation has its costs and benefits. While the costs certainly result from potential under-regulation and the according welfare losses from market power, the benefits stem from the avoidance of a heavy-handed regulatory apparatus and greater flexibility for firms. A critical issue that has been identified is the credibility of the regulatory threat. If the threat of regulation is not credible, a LHRR amounts to leaving monopolistic firms unregulated, clearly an undesirable situation from a welfare point of view.

While the light-handed approaches to regulation in the German electricity industry have been almost completely abandoned, New Zealand's telecommunications regulation has undergone a less drastic shift with the establishment of a sector-specific regulator within the Commerce Commission. With respect to New Zealand's electricity industry, it remains to be seen as to how the regulatory regime will develop in the future. So far, the innovative threshold regime has not been adopted.

5 Summary and Conclusions

While utility and other network industries have undergone significant reforms over the last 20 years characterised by processes of privatisation, liberalisation and de- and re-regulation, there is still a vivid and ongoing debate about the proper balance between regulation and competition policy today. The focus of this discussion is on the questions of (a) how over-regulation can be avoided and (b) how and under what circumstances intrusive heavy-handed sector-specific regulation can be substituted by reliance on competition law or at least a more light-handed approach.

In designing an "optimal" regulatory regime, there are usually five broad objectives Governments look for when seeking to regulate network industries:

- Constraining companies from earning monopoly rents (allocative efficiency),
- maintaining incentives for network operators to seek productive efficiency,
- maintaining incentives for both operators and consumers to make optimal investment and consumption decisions (dynamic efficiency),
- minimising the administrative and compliance costs associated with the regulatory regime (transaction cost efficiency),
- minimising the extent to which the regime creates an incentive for lobbying, i.e., directly unproductive rent seeking (political efficiency), where transaction cost efficiency and political efficiency may be summarised as institutional efficiency.

The difficulty is that no regulatory regime is perfect. In general, there are two risks or types of errors associated with regulatory policy. A type-I error occurs if a market or firm is regulated intensely while it should be subject to a more light-handed approach, while a type-II error occurs if the regulatory framework leaves a firm too softly regulated while it should be regulated more intensely. Given the political economy of regulatory agencies, it is well accepted that there is a risk of over-regulation; put differently, regulatory agencies tend to place too much emphasis on reducing type-II errors while they tend to underestimate the risk of type-I errors.

In order to avoid over-regulation, some countries have experimented with what might be called "threat-based regulation". The idea is - similarly to the contestable market model - that *actual* regulation may not be necessary as long as the threat of regulation is credible. In that case, so the idea, even monopolies will abstain from setting monopolistic prices in order to avoid being regulated.

Threat-based regulation on which a light-handed regulatory regime is usually built may be regarded as one of many alternative institutional frameworks to govern network industries. Among the numerous possible institutional frameworks, threat-based regulation is situated between pure reliance on general competition law, which is usually applied ex post (with the notable exemption of merger control), at the one end and sector-specific ex ante regulation at the other end of the spectrum.

From a new institutional economics perspective, a LHHR can be viewed as a relational, administered, (and necessarily incomplete) contract. While the essential facilities' owners implicitly promise to provide access to their infrastructure on fair and non-discriminatory terms, consumers promise not to hold up investors by requiring (by means of the Government) price cuts once the firms have undertaken specific investments. Since it would be too costly, however, to contractually fix all possible contingencies ex ante (due to the contracting parties' bounded rationality), the contract will remain necessarily incomplete. The contracting parties will rather explicitly or implicitly agree upon contractual clauses or a "constitution of the contract" by which changes in future conditions lead to renegotiations. Since a third party (such as a sector-specific regulatory authority, a competition authority or an independent court) acts as an "arbitrator", one usually speaks of an „*administered contract*" (see Goldberg, 1976).

However, relational contracts cannot fully be enforced through third parties. Therefore, they need to be self-enforcing in order to successfully govern relationships. For a contract to be self-enforcing though, there must be *credible* threats if either of the contracting parties deviates from the "spirit of the contract".

As experience from Germany and New Zealand has demonstrated, there are quite a number of problems associated with threat-based regulatory systems. The most significant problem has turned out to be the likely under-regulation and according welfare losses from market power, and in this context, a critical issue that has been identified is the credibility of the regulatory threat. If the threat of regulation is not credible, a LHRR amounts to leaving monopolistic firms unregulated, clearly an undesirable situation from a welfare point of view.

This failure of threat-based regulatory systems results from potential free-rider problems associated with good corporate behaviour (as in the case of Germany's electricity market) as well as the lack of credibility of the threat either because of the weakness of the enforcement authorities or due to political constraints. Furthermore, a problem emerges if exercising the threat of regulation implies an abandonment of the LHRR and a corresponding substitution with heavy-handed ex ante regulation.

An innovative way out of this dilemma has been the quasi-automatic threshold system originally designed for New Zealand's electricity distribution networks, which aims at simulating the pressures that companies would face in competitive markets. While this innovative system has been discussed intensively in New Zealand, it has not yet been implemented. Instead, there is still an ongoing debate about the appropriate thresholds and mechanisms, and the Government has installed a rather toothless beast – the Electricity Commission – to supervise the electricity sector while lacking necessary powers. Hence, it remains to be seen how this hybrid model might develop.

In summary, the models of light-handed regulation that have been used thus far have unfortunately not lived up to their promises so that further regulatory innovations are required to make a LHRR work in practice.

Appendix

Table 1. Example of the option for setting the thresholds for price control (WACC is assumed to be 8%) – taken from the Ministry of Commerce (1998)

Price Performance Measure for Line Businesses

Firm	Price (¢/kWh)	Ranking	Calculation	Demerit points
Lineco1	11.00	5	5/9•100	55.56
Lineco2	10.00	2	2/9•100	22.22
Lineco3	11.75	7	7/9•100	77.78
Lineco4	10.63	4	4/9•100	44.44
Lineco5	11.38	6	6/9•100	66.67
Lineco6	8.50	0	0/9•100	0.00
Lineco7	12.50	9	9/9•100	100.00
Lineco8	12.13	8	8/9•100	88.89
Lineco9	10.13	3	3/9•100	33.33
Lineco10	9.50	1	1/9•100	11.11

Rate of Return (ROR) Performance Measure for Line Businesses

Firm	ROR (%)	Ranking	Calculation	Demerit points
Lineco1	4.4	0	0/4•100	0.00
Lineco2	9.5	3	3/4•100	75.00
Lineco3	8.1	1	1/4•100	25.00
Lineco4	5	0	0/4•100	0.00
Lineco5	9	2	2/4•100	50.00
Lineco6	0	0	0/4•100	0.00
Lineco7	14	4	4/4•100	100.00
Lineco8	-5	0	0/4•100	0.00
Lineco9	1	0	0/4•100	0.00
Lineco10	6	0	0/4•100	0.00

Service Quality Performance Measures for Line Businesses

Firm	Service Quality	Ranking	Calculation	Demerit points
Lineco1	200	9	9/9•100	100.00
Lineco2	90	3	3/9•100	33.33
Lineco3	95	4	4/9•100	44.44
Lineco4	20	0	0/9•100	0.00
Lineco5	175	8	8/9•100	88.89
Lineco6	100	5	5/9•100	55.56
Lineco7	80	2	2/9•100	22.22
Lineco8	150	7	7/9•100	77.78
Lineco9	100	5	5/9•100	55.56
Lineco10	20	0	0/9•100	0.00

Total Demerit Points for Line Businesses

		Weighted Demerit Points		
Firm	Price	ROR	Service Quality	Total
Lineco1	55.56•0.6=33.33	0.00•0.1=0.00	100.00•0.3=30.00	63.33
Lineco2	22.22•0.4=8.89	75.00•0.3=22.50	33.33•0.3=10.00	41.39
Lineco3	77.78•0.6=46.67	25.00•0.1=2.50	44.44•0.3=13.33	62.50
Lineco4	44.44•0.6=26.67	0.00•0.1=0.00	0.00•0.3=0.00	26.67
Lineco5	66.67•0.6=40.00	50.00•0.1=5.00	88.89•0.3=26.67	71.67
Lineco6	0.00•0.6=0.00	0.00•0.1=0.00	55.56•0.3=16.67	16.67
Lineco7	100.00•0.6=60.00	100.00•0.1=10.00	22.22•0.3=6.67	76.67
Lineco8	88.89•0.6=53.33	0.00•0.1=0.00	77.78•0.3=23.33	76.67
Lineco9	33.33•0.6=20.00	0.00•0.1=0.00	55.56•0.3=16.67	36.67
Lineco10	11.11•0.6=6.66	0.00•0.1=0.00	0.00•0.3=0.00	6.66

Points to note from the table:

- The maximum weighted-demerit points that a line business can incur for: (i) price is 60 (Lineco7); (ii) ARP is 30 (no line business has incurred the maximum weighted-demerit points for ARP - This is because Lineco7, which has the highest ARP, was given a 0.1 weight for ARP, rather than 0.3); and (iii) service quality 30 (Lineco1).

the highest ARP, was given a 0.1 weight for ARP, rather than 0.3); and (iii) service quality 30 (Lineco1).

- The minimum possible total (weighted) demerit points is 0, and the maximum is 100. None of the line businesses matched these values. The closest to the minimum was Lineco10 with 6.66 demerit points. The closest to the maximum were Lineco7 and Lineco8, each with 76.67, respectively.
- Only Lineco2's demerit points were based on the weightings a = 0.4, b = 0.3 and c = 0.3. This reflects that Lineco2's price ranking was very favourable relative to its rate of return ranking.
- Only Lineco7 and Lineco8 had 75 or more demerit points. Lineco8 exceeded the 75 point threshold despite having and ARP of -5%. Given the current level of prices, Lineco8 could, for example, reduce its demerit points below 75 in the future by decreasing its prices by 0.37¢/kWh.
- Lineco1, Lineco3 and Lineco5 would have to reduce their prices, profits and/or improve their service quality if they want to avoid the risk of breaching the threshold in the future (as other line businesses improve their relative position). If, for example, Lineco2 and Lineco7 dropped their respective ARPs by 0.6% and 5.1%, Lineco5's demerit points would exceed 75 in the following year.
- Similarly, Lineco5 could increase its prices by no more than 0.375¢/kWh without exceeding the 75 threshold. Even if Lineco5 increased its prices by less than 0.25¢/kWh it would still exceed the threshold, if this resulted in its ARP increasing by 5.1%.

Source: Ministry of Commerce (1998)

References

ABREU, D., PEARCE, D. and STACCHETTI, E. (1984), Optimal Cartel Equilibria with Imperfect Monitoring, *Journal of Economic Theory*, 39, 251-69.

ACUTT, M. and ELLIOT, C. (2001a), Threat-Based Competition Policy, *European Journal of Law and Economics*, 11, 309-17.

ACUTT, M. and ELLIOT, C. (2001b), *Threat-Based Regulation and Endogenously Determined Punishment*, Working Paper, Lancaster University Management School, 2001/007.

AREEDA, P. (1990), Essential Facilities: An Epithet In Need Of Limiting Principles, *Antitrust Law Journal*, 58, 841-53.

BALDWIN, R. and CAVE, M. (1999), *Understanding Regulation, Theory, Strategy, and Practice*, Oxford University Press: Oxford.

BARZEL, Y. (1989), *Economic Analysis of Property Rights*, Cambridge University Press: Cambridge

BAUMOL, W.J. and SIDAK, G.J. (1994), The Pricing of Inputs Sold to Competitors, *Yale Journal on Regulation*, 11, 171-202.

BIER, C. (2002), *Regulierter oder Verhandelter Zugang zum Stromnetz? Eine ökonomische Analyse unter Berücksichtigung imperfekter Aufsichtsbehörden*, Verlag Kovac: Hamburg.

BLANCHARD, C. (1994), Telecommunications Regulation in New Zealand: How Effective is' Light-Handed Regulation?, *Telecommunications Policy*, 18, 154-64.

BLANCHARD, C. (1995), Telecommunications Regulation in New Zealand: Light-Handed Regulation and the Privy Council's Judgement, *Telecommunications Policy*, 19, 465-75.

BOLLARD, A.E. and PICKFORD, M. (1996), *Utility Regulation in New Zealand*, Wellington.

BOLLARD, A.E. and PICKFORD, M. (1997), Utility Regulation in New Zealand, in BEESLEY, M.E. (ed.): *Regulating Utilities: Broadening the Debate*, Institute for Economics Affairs: London, 75-131.

BRUNEKREEFT, G. (2003), *Regulation and Competition Policy in the Electricity Market: Economic Analysis and German Experience*, Nomos Verlag: Baden-Baden.

BRUNEKREEFT, G. (2004), Regulatory Threat in Vertical Related Markets: The Case of German Electricity, *European Journal of Law end Economics*, 17, 285-305.

BRUNEKREEFT, G. and TWELEMANN, S. (2005), Regulation, Competition and Investment in the German Electricity Market: RegTP or REGTP, *Energy Journal* 26, Special Issue on European Electricity Liberalisation, 99-126.

CABINET OFFICE (2005), *Cabinet Minute of Decision: Control of Gas Pipeline Services: Policy Approval and Order in Council*, CAB Min (05) 27/3, 25 July 2005, Cabinet Office: Wellington.

COMMERCE COMMISSION (2003), *Regulation of Electricity Lines Businesses: Targeted Control Regime*, Commerce Commission: Wellington.

COMMERCE COMMISSION (2004), *Regulation of Electricity Lines Businesses: Targeted Control Regime*, *Threshold Decisions*, Commerce Commission: Wellington.

CROCKER, K.J. and MASTEN, S.E. (1996), Regulation and Administered Contracts Revisited: Lessons from Transaction-Cost Economics for Public Utility Regulation, *Journal of Regulatory Economics*, 9, 5-39.

DIXIT, A.K. (1996), *The Making of Economic Policy: A Transaction-Cost Politics Perspective*, MIT Press: Cambridge, MA.

EVANS, L., GRIMES, A., WILKINSON, B. and TEECE, D. (1996), Economic Reform in New Zealand 1984-95: The Pursuit of Efficiency, *Journal of Economic Literature*, 34, 1856-902.

GALT, D. (1998), Telecommunications, in: BOLLARD, A. and PICKFORD, M. (eds.), *The Structure and Dynamics of New Zealand Industries*, Dunmore Press: Palmerston North, 221-54.

GLAZER, A. and MCMILLAN, H. (1992), Pricing by the Firm under Regulatory Threat, *Quarterly Journal of Economics*, 107, 1089-99.

GOLDBERG, V.P. (1976), Regulation and Administered Contracts, *Bell Journal of Economics*, 7, 427-48.

GREEN, E.J. and PORTER, R.H. (1984), Noncooperative Collusion under Imperfect Price Information, *Econometrica*, 52, 87-100.

HAUCAP, J. (2003), Wettbewerb und Regulierung im Mobilfunk aus Sicht der ökonomischen Theorie, in: KRUSE, J. and HAUCAP, J. (eds.): *Mobilfunk zwischen Wettbewerb und Regulierung*, Fischer: München, 47-76.

HAUCAP, J. and KRUSE, J. (2004), Ex-Ante-Regulierung oder Ex-Post-Aufsicht für netzgebundene Industrien?, *Wirtschaft und Wettbewerb*, 3, 266-75.

HAYEK, F.A. von (1976), Die Verwertung des Wissens in der Gesellschaft, in: HAYEK, F.A. von (ed.): *Individualismus und wirtschaftliche Ordnung*, Salzburg, 103-21.

KING, S.P. and MADDOCK, R. (1999), Light-Handed Regulation of Access in Australia: Negotiation with Arbitration, *Information Economics and Policy*, 11, 1-22.

KNIEPS, G. (1999), Deregulierung und die Dynamik des Wettbewerbs in der Telekommunikation, in: OBERENDER, P. (ed.): *Die Dynamik der Telekommunikationsmärkte als Herausforderung an die Wettbewerbspolitik*, Duncker und Humblot: Berlin, 9-14.

KNIEPS, G. (2001), Zur Regulierung monopolistischer Bottlenecks, *Zeitschrift für Wirtschaftspolitik*, 48, 297-304.

KNIEPS, G. and BRUNEKREEFT, G. (2003), *Zwischen Regulierung und Wettbewerb – Netzsektoren in Deutschland*, Physica: Heidelberg.

KRUSE, J. (2001), Entgeltregulierung bei Fern- und Auslandsverbindungen?, in: IMMENGA, U., KIRCHNER, C., KNIEPS, G. and KRUSE, J. (eds.): *Telekommunikation im Wettbewerb. Eine ordnungspolitische Konzeption nach drei Jahren Marktöffnung*, C.H. Beck: München, 111-27.

KUMKAR, L. (1997), *Nach der Verabschiedung der Stromrichtlinie: Status quo und Perspektiven der europäischen Elektrizitätswirtschaft*, Institut für Weltwirtschaft: Kiel.

LAFFONT, J.J. and TIROLE, J. (2000), *Competition in Telecommunications*, MIT Press: Cambridge, MA.

LAFFONT, J.J. and TIROLE, J. (2003), *A Theory of Incentives in Procurement and Regulation*, MIT Press: Cambridge, MA.

LEIBENSTEIN, H. (1966), Allocative Efficiency vs. X-Efficiency, *American Economic Review* 56, 392-415.

LEVY, B. and SPILLER, P. (1994), The Institutional Foundations of Regulatory Commitment: A Comparative Analysis of Telecommunications Regulation, *Journal of Law, Economics, and Organization*, 10, 201-46.

LIPSKY, A.B. and SIDAK, G.J. (1999), Essential Facilities, *Stanford Law Review*, 51, 1187-1248.

LITTLECHILD, S. (2003), Electricity: Regulatory Developments Around the World, in: ROBINSON, C. (ed.): *Competition and Regulation in Utility Markets*, Edward Elgar: Cheltenham, 61-87.

LUXTON, J., WILLIAMSON, M. and KIDD, D. (1996), *Government Signals Future Directions for Regulation of Telecommunications, Electricity and Gas*, Wellington.

MACNEIL, J.R. (1974), The Many Futures of Contract, *Southern California Law Review*, 47, 691-816.

MINISTRY OF COMMERCE (1995), *Light-Handed Regulation of New Zealand's Gas Industries*, Ministry of Commerce: Wellington.

MINISTRY OF COMMERCE (1998), *Electricity Reform: Discussion Paper on the Operation of the Specific Thresholds for Price Control for Electricity Line Businesses*, Ministry of Commerce Energy Markets Policy Group: Wellington.

MINISTRY OF COMMERCE (1999), *Proposals for the Operation of the Specific Thresholds for Price Control for Electricity Line Businesses*, Ministry of Commerce Energy Markets Policy Group: Wellington.

MINISTRY OF COMMERCE/THE TREASURY (1995), *Regulation of Access to Vertically-integrated Natural Monopolies – A Discussion Paper*, Ministry of Commerce/The Treasury: Wellington.

MONOPOLKOMMISSION (2002), *Netzwettbewerb durch Regulierung*, 14. Hauptgutachten, Nomos Verlag: Baden-Baden.

MONOPOLKOMMISSION (2004), *Wettbewerbspolitik im Schatten „Nationaler Champions"*, 15. Hauptgutachten, Nomos Verlag: Baden-Baden.

NORTH, D.C. (1990), *Institutions, Institutional Change and Economic Performance*, Cambridge University Press: Cambridge.

OECD (1999), *Relationship between Regulators and Competition Authorities*, OECD: Paris.

OFTEL (2000), *The Application of the Competition Act in the Telecommunications Sector*, Oftel: London.

PANZAR, J.C. and WILLIG, R.D. (1981), Economies of Scope, *American Economic Review*, 71, 268-72.

PANZAR, J.C. (1989), Technological Determinants of Firm and Industry Structure, in SCHMALENSEE, R. and WILLIG, R.D. (eds.): *Handbook of Industrial Organization*, Vol. I, North Holland: Amsterdam, 3-59.

PELTZMAN, S. (1976), Toward a More General Theory of Regulation, *Journal of Law and Economics,* 19, 211-40.

PICKFORD, M. (1996), Information Disclosure, the Baumol-Willig Rule, and the 'Light-Handed' Regulation of Utilities in New Zealand, *New Zealand Economic Papers*, 30, 199-218.

POSNER, R.A. (1974), Theories of Economic Regulation, *Bell Journal of Economics and Management Science*, 5, 335-58.

RICHTER, R. and FURUBOTN, E.G. (2003), *Neue Institutionenökonomik: Eine Einführung und kritische Würdigung*, Mohr Siebeck: Tübingen.

ROTTENBILLER, S. (2002), *Essential Facilities als ordnungspolitisches Problem*, Verlag Peter Lang: Frankfurt am Main.

SCHEBSTADT, A. (2004): Die Aufsicht über Netznutzungsentgelte zwischen Kartellrecht und sektorspezifischer Regulierung, *Recht der Energiewirtschaft*, 8, 181-208.

SIDAK, J. and SPULBER, D.F. (1997), *Deregulatory Takings and the Regulatory Contract*, Cambridge University Press: Cambridge.

STIGLER, G. (1971), The Economic Theory of Regulation, *Bell Journal of Economics and Management Science*, 2, 3-21.

TRAIN, K.E. (1991), *Optimal Regulation: The Economic Theory of Natural Monopoly*, MIT Press: Cambridge, MA.

TULLOCK, G. (1967), The Welfare Cost of Tariffs, Monopolies, and Theft, *Western Economic Journal*, 5, 224-32.

TYE, W.B. and LAPUERTA, C. (1996), The Economics of Pricing Network Interconnection: Theory and Application to the Market for Telecommunications in New Zealand, *Yale Journal on Regulation*, 13, 419-500.

VDN (2005), *Netzzugangsentgelte der deutschen Netzbetreiber*, VDN: Berlin (www.vdn-berlin.de)

VICKERS, J. (2005), Abuse of Market Power, *Economic Journal*, 115, 244-61.

VOGELSANG, I. (2003), The German Telecommunications Reform – Where Did it Come From, Where is it, and Where is it Going?, *Perspektiven der Wirtschaftspolitik* 4, 313-40.

WEBB, M. and TAYLOR, M. (1998), Light-Handed Regulation of Telecommunications in New Zealand: Is Generic Competition Law Sufficient?, *International Journal of Communications Law and Policy*, 2, 1-17.

WETZEL, D. (2003a), Sonderweg der Stromwirtschaft droht in der Sackgasse zu enden, *Die Welt*, 2. April 2003.

WETZEL, D. (2003b), Harte Eingriffe in den Strommarkt, *Die Welt*, 29. August 2003.

WEIZSÄCKER, C.C. von (1982), Staatliche Regulierung - positive und normative Theorie, *Schweizerische Zeitschrift für Volkswirtschaft und Statistik*, 118, 325-43.

WELSON, E. and THOMSON, T. (2002), Reform of the New Zealand Electricity Industry: 15 Years and Still Counting, *Utilities Law Review*, 12:5, 115-119.

WILLIAMSON, B. and MUMSSEN, Y. (2000), *Economic Regulation of Network Industries*, Working Paper No. 5, New Zealand Treasury: Wellington.

WILLIAMSON, O.E. (1996), *The Economic Institutions of Capitalism*, The Free Press: New York.

WILLIAMSON, O.E. (1996), *The Mechanisms of Governance*, Oxford University Press: Oxford.

Information Society 2015: Towards Increased Participation

Jaroslaw K. Ponder[1]

1 Introduction 203
2 Global Trends in the Telecommunication Sector 203
3 The Digital Divide and the WSIS 206
4 Conclusions 209

1 Introduction

The relevance of ICTs for socio-economic development is becoming more and more significant for the developing countries, as well as for economies in transition. Considering the tendency of developed countries to transform from the traditional structured economy, based mainly on work and capital, into economies based on knowledge (where the ICTs play a fundamental role), the process of building an open, inclusive and global Information Society (IS) has a special meaning. Particularly in these times of the progressing globalization process, access to the borderless virtual world may create opportunities for bridging socio-economic divides.

Despite global efforts and intensive reforms in telecommunications and the ICT sector worldwide, for many countries the problem of the Digital Divide remains very real. Disparities in access to the ICTs between, on the one hand developing countries and economies in transition, and on the other the developed world, are still evident. However we observe the continuous decline in differences, the catch-up process can take many years. The recent World Summit on the Information Society (WSIS) – concluded in November 2005 in Tunis – achieved a global understanding on the complexity of the development process of the Information Society and, among its many other decisions, set short, medium and long term goals, as well as actions to promote and facilitate the increased participation of all stakeholders in the Information Society.

2 Global Trends in the Telecommunication Sector

Over the last two decades, the traditional telecommunication sector has changed significantly. Since the publication of the ITU's Maitland report in 1985, in which the authors underlined the importance of telecommunications for development, the

---

[1] The views expressed in this article are those of the author and do not necessarily reflect the opinions of the ITU.

phenomenon of the Digital Divide has become much more complex. In fact, the recently observed broad popularization of wireless technologies and the Internet has thoroughly altered both the meaning and forms of communication. Despite continuous growth in the number of fixed telephone lines over a long period, since 2002, the number of cellular phones has exceeded the amount of fixed lines worldwide. Within one decade, mobile telephones spread to such extent that in 2005, they exceeded two billion worldwide (see Figure 1). In addition, the number of Internet users has been growing very fast recently and stands, at present, close to one billion. Very strong network externalities continuously boost demand for access to the Internet and are setting new standards, in terms of broadband connectivity, and simultaneously creating new challenges for the Digital Divide.

The rapid popularization of the communication technologies owes its high growth dynamics to many factors. One of the most remarkable is the regulatory reform that, within a very short time, has reinforced the role of natural market mechanisms in the fight against the Digital Divide. The broadly promoted, pro-competitive telecommunication policy has caused the overall size of the telecommunication and ICT sector to grow fast and investment in telecommunication infrastructure worldwide has been increasing constantly over the last ten years. Many newcomers have entered the sector and have been extending their activities continuously. The overall telecommunication and ICT innovation dynamics have become one of the highest among all the economic sectors. As a consequence of this development, the relative costs of telecommunication equipment have been falling continuously, even as its functionality has been improving. In line with the liberalization process, the relative prices of communication services have decreased significantly. At the moment, this positive tendency is being strengthened through the general trend towards migration into the IP environment, which enables traditional services to be offered at minimal prices; this partially marginalizes the problem of affordability. Despite these trends, the problem of the digital divide still exists and, for many countries, is very difficult to solve. Without clearly defined public policy to promote particular concepts, such as universal access or universal service, it might be difficult to achieve the satisfactory level of connectivity that guarantees increased participation in the Information Society (see Figure 1). Taking into consideration national particularities, the goals and measures of public policy may be very different in comparison to the international agenda.

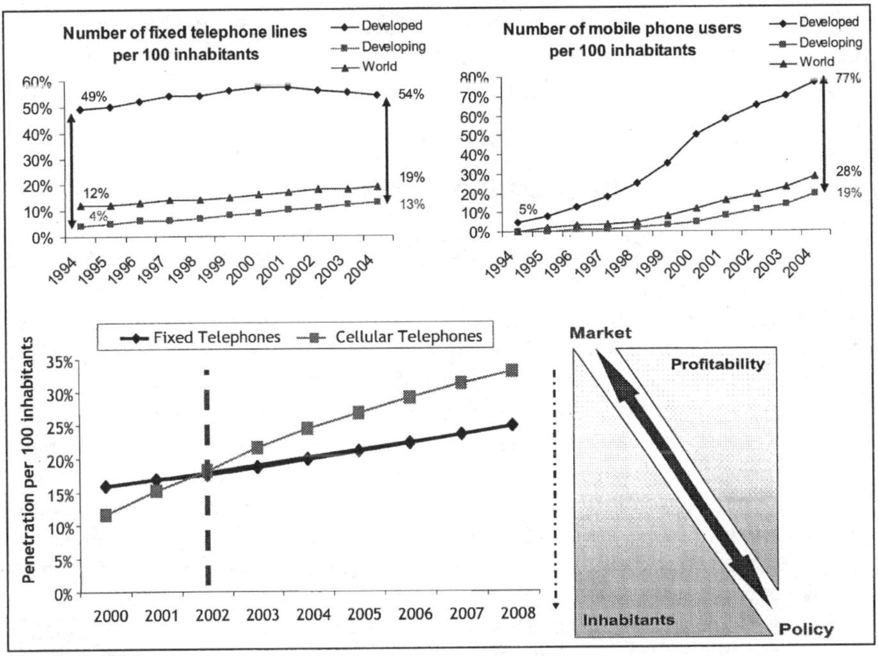

Source: ITU

Fig. 1. Trends in penetration rates for fixed and mobile telecommunications worldwide and regulatory concerns on universal access

When talking about global trends in the telecommunication and ICT sectors, it is important to stress that the evolving process of convergence is becoming much more meaningful than it has ever been and, together with the gradual deployment of IP-enabled next generation networks (NGNs), this tendency continues to grow. The IP environment offers much greater flexibility in the provision of diverse services, e.g. voice, video and data thanks to one network, the setting up and maintenance of which may be significantly cheaper in comparison to the existing solutions. Gains of up to 40 per cent in CAPEX and OPEX have increased expectations that the cost of service provision may be reduced in relative terms, thus making it more affordable, particularly for poor countries. Particular attention should be drawn to the potential of next generation wireless access technologies, such as WiFi and WiMAX, which may become effective tools in the fight against any lack of telecommunication infrastructure. India's wide deployment of wireless technologies provides one of the best examples of the deployment of these technologies as a backbone infrastructure.

These few observations pointed out above demonstrate that the telecommunication and ICT sector is expanding very fast and, in fact, is becoming a backbone for each economy. ICTs create many opportunities, but also many challenges, in particular for poor countries, which have to find a way to take the best possible benefit from this innovation for the socio-economic catch-up process. There is no single, unified recipe for all, but a key to the success is a global understanding of possible steps at the national, regional and international level that would facilitate the development of the Information Society, which amongst many others things, has been achieved during the World Summit on the Information Society.

3 The Digital Divide and the WSIS

The WSIS has gained a special and permanent place in the history of the Information Society. Formally, it was started in 1998 by the International Telecommunication Union and since 2001, in accordance with United Nations Resolution 56/183, became a UN Summit. Given the complexity of the issues discussed, and its proposed multi-stakeholder character, it was decided to hold it in two phases. The preparations had to take place through an open-ended, intergovernmental Preparatory Committee, the tasks of which included defining the agenda of the Summit, deciding on the modalities of the participation of other stakeholders in the Summit, and finalizing both the draft Declaration of Principles and the draft Plan of Action. Consequently, the WSIS process was accompanied by numerous official preparatory meetings, ministerial conferences, thematic meetings, as well as many parallel events.

In order to ensure its true multi-stakeholder character, all parties involved in the development of the Information Society, i.e. governments, relevant UN bodies, international organizations, the private sector and civil society, were strongly encouraged, at the national, regional and global level, to participate pro-actively in both the preparatory process and the Summit itself. Over time, the interest in active participation in the process has grown.

Basically, the objective of the first phase of the WSIS, the so-called Geneva Phase, was to develop and foster a clear statement of political will and take concrete steps to establish the foundations for an Information Society for all, that reflected all the different interests at stake. The objective of the second phase, the so-called Tunis phase, was to put Geneva's Plan of Action into motion, as well as to find solutions and reach agreements in the fields of Internet governance, financing mechanisms, and follow-up and implementation of the Geneva and Tunis documents. In other words, the second phase of WSIS became the Summit of Solution, whereas the first phase provided the roadmap of where to go and how to get there.

The final outcome of the World Summit on the Information Society includes four documents that may be considered in two sub-groups. In the Geneva Declaration and Tunis Commitment, stakeholders shared general vision and principles of overcoming the digital divide and achieving Millennium Development Goals. They affirmed a unified approach to cybersecurity and reiterated the commitment to freedom of expression online. The involved parties agreed on the need to create an enabling environment based on principles like good governance, transparency or a pro-competitive policy approach. Moreover, they encouraged the development of localized content and multilingualism.

The character of two other outcome documents, i.e. the Geneva Plan of Action and the Tunis Agenda for the Information Society, is slightly different. In these documents, the stakeholders confirm the will to build an inclusive and development-oriented Information Society and to determine the particular means to achieve this. Consequently, eleven action lines have been defined (see Box 1). According to these definitions, the Information Society should be monitored and its development process moderated.

These documents also set specific, concrete targets to be achieved by 2015. In terms of countering of the phenomenon of the Digital Divide, countries committed, by all means, to enhance connectivity and universal, ubiquitous, equitable, non-discriminatory and affordable access to, and use of, ICTs, taking into account their different national circumstances. Stakeholders also confirmed that the ICTs should be also used as a tool to achieve the internationally agreed development goals and objectives, including the Millennium Development Goals. More concretely, in accordance with the documents, by 2015, the population of the entire world should have access to television and radio services and half of world's inhabitants should have access to ICTs.

The involved parties agreed that the process should be monitored and evaluated regularly. For this purpose, the ITU-led WSIS Stocktaking exercise should be continued. The aim of this exercise should be to build and maintain an inventory of the activities undertaken by governments and all stakeholders in implementing all the decisions included in the outcome documents. The overall review and evaluation of the advancements in the development of the information society should take place in 2015.

In addition, the stakeholders confirmed the importance of ICT indicators in measuring the Digital Divide and endorsed the Partnership on Measuring ICT for Development, as well as two indices, i.e. the ICT Opportunity Index and the Digital Opportunity Index built upon the common set of core indicators. Also, the role of the multi-stakeholder partnerships for the development of an open and inclusive Information Society has been underlined. The ITU-led Connect the World initiative should serve as the best example of a platform that enables the partners to interact and join forces in countering the Digital Divide.

The issues of financing of ICTs for development and Internet governance were discussed specifically during the Tunis phase. Therefore, the Tunis Agenda has, among other things, addressed the issue of the Digital Solidarity Fund, financing mechanisms encouraging development of infrastructure or the creation of a new Internet Governance Forum, which will carry forward the work on Internet Governance.

In the Tunis Agenda, the stakeholders also reaffirmed their dedication to the foundations of the Information Society and outlined the basis for its implementation and follow-up at the national, regional and international levels. The multi-stakeholder character should be preserved at each level. Nevertheless, as regards its implementation at the international level, it is important to stress the distinction between the UN coordination process that will follow the UN mechanisms and the multi-stakeholder implementation. To meet the needs of the latter, a list of possible facilitators and moderators for all eleven action lines has been created. In the meantime, the consultative process – which took place directly after the Summit – enabled the appointment of the focal points for each action line and the extension of the list by adding a few international organizations and civil society stakeholders (see Box 1).

Box 1 WSIS action lines and their facilitators as well as co-facilitators

Action Line	Facilitators
C1. The role of public governance authorities and all stakeholders in the promotion of ICTs for development	ECOSOC/UN Regional Commissions/ITU/[UN DESA]
C2. Information and communication infrastructure	ITU/[APC]
C3. Access to information and knowledge	ITU/UNESCO/[FAO/UNIDO]
C4. Capacity building	UNDP/UNESCO/ITU/UNCTAD/[UN DESA/FAO/UNIDO]
C5. Building confidence and security in the use of ICTs	ITU
C6. Enabling environment	ITU/UNDP/UN REGIONAL COMMISSIONS/UNCTAD/[UN DESA/UNIDO/APC]
C7. ICT Applications	
E-government	[UN DESA]/UNDP/ITU
E-business	WTO/UNCTAD/ITU/UPU
E-learning	UNESCO/ITU/UNIDO
E-health	WHO/ITU
E-employment	ILO/ITU
E-environment	WHO/WMO/UNEP/UN- Habitat/ITU/ICAO
E-agriculture	FAO/ITU
E-science	UNESCO/ITU/UNCTAD/[WHO]
C8. Cultural diversity and identity, linguistic diversity and local content	UNESCO
C9. Media	UNESCO
C10. Ethical dimensions of the Information Society	UNESCO/ECOSOC/[WHO/ECPAT Int'l]
C11. International and regional cooperation	UN Regional Commissions/ UNDP/ITU/UNESCO/ECOSOC/[UN DESA]

Note : The list takes into account the decisions made during multi-stakeholder consultation process that took place after the Summit in 2006. Civil society entities are indicated in italics. Those agencies shown in bold are the provisional focal points for each action line.
Source: ITU

4 Conclusions

The development of the Information Society should be seen as an on-going process that, particularly in times of progressing globalization, is becoming much more meaningful than it has ever been before. The World Summit on the Information Society was definitely a success and its consequences will be noticeable in the short, medium and long term. The WSIS confirmed the political will to foster the further development of the Information Society. Nevertheless, nowadays it is the role of all stakeholders to make the WSIS implementation real at the national, regional and international levels, in order to establish solid links between the borderless, virtual world and the real one, without leaving anyone on the sidelines.

List of Contributors

WOLFGANG BRIGLAUER
Austrian Regulatory Authority for Broadcasting and Telecommunications (RTR), Economics Division

FRIEDHELM DOMMERMUTH
Department "Economic Aspects of Telecommunications Regulation" of the Regulatory Authority for Telecommunications and Posts (Reg TP)

CHRIS DOYLE
Warwick Business School, Warwick University

MORTEN FALCH
Center for Information and Communication Technologies, Technical University of Denmark

JUSTUS HAUCAP
Ruhr-University of Bochum, Department of Economics

ULRICH HEIMESHOFF
Ruhr-University of Bochum, Department of Economics

ANDERS HENTEN
Center for Information and Communication Technologies, Technical University of Denmark

DOMINIK KUROPKA
Hasso Plattner Institute of IT Systems Engineering at University of Potsdam

JAROSLAW K. PONDER
International Telecommunication Union, Geneva

RICHARD TEE
TNO - Information and Communication Technology Department ICT and Policy Studies on Media, Information and Telecommunication (SMIT) - IBBT at the University of Brussels (VUB)

ANDRÉ UHDE
Ruhr-University of Bochum, Department of Economics

PAUL J.J. WELFENS
European Institute for International Economic Relations (EIIW) at the University of Wuppertal

MATHIAS WESKE
Hasso Plattner Institute of IT Systems Engineering at University of Potsdam

Further Publications by *Paul J. J. Welfens*

M. W. Klein, P. J. J. Welfens (Eds.)
Multinationals in the New Europe and Global Trade
1992. XV, 281 pages. 24 Figs., 75 Tab.,
Hardcover, ISBN 3-540-54634-0

P. J. J. Welfens (Ed.)
Economic Aspects of German Unification
Expectations, Transition Dynamics and International Perspectives
1996. XV, 527 pages. 34 Figs., 110 Tab.,
Hardcover, ISBN 3-540-60261-5

R. Tilly, P. J. J. Welfens (Eds.)
European Economic Integration as a Challenge to Industry and Government
Contemporary and Historical Perspectives on International Economic Dynamics
1996. X, 558 pages. 43 Figs.,
Hardcover, ISBN 3-540-60431-6

P. J. J. Welfens (Ed.)
Economic Aspects of German Unification
Expectations, Transition Dynamics and International Perspectives
2nd revised and enlarged edition
1996. XV, 527 pages. 34 Figs., 110 Tab.,
Hardcover, ISBN 3-540-60261-5

P. J. J. Welfens
European Monetary Integration
EMS Developments and International Post-Maastricht Perspectives
3rd revised and enlarged edition
1996. XVIII, 384 pages. 14 Figs., 26 Tab.,
Hardcover, ISBN 3-540-60260-7

P. J. J. Welfens (Ed.)
European Monetary Union
Transition, International Impact and Policy Options
1997. X, 467 pages. 50 Figs., 31 Tab.,
Hardcover, ISBN 3-540-63305-7

P. J. J. Welfens, G. Yarrow (Eds.)
Telecommunications and Energy in Systemic Transformation
International Dynamics, Deregulation and Adjustment in Network Industries
1997. XII, 501 pages. 39 Figs.,
Hardcover, ISBN 3-540-61586-5

P. J. J. Welfens, H. C. Wolf (Eds.)
Banking, International Capital Flows and Growth in Europe
Financial Markets, Savings and Monetary Integration in a World with Uncertain Convergence
1997. XIV, 458 pages. 22 Figs., 63 Tab.,
Hardcover, ISBN 3-540-63192-5

P. J. J. Welfens, D. Audretsch, J. T. Addison, H. Grupp
Technological Competition, Employment and Innovation Policies in OECD Countries
1998. VI, 231 pages. 16 Figs., 20 Tab.,
Hardcover, ISBN 3-540-63439-8

P. J. J. Welfens, G. Yarrow, R. Grinberg, C. Graack (Eds.)
Towards Competition in Network Industries
Telecommunications, Energy and Transportation in Europe and Russia
1999. XXII, 570 pages. 63 Figs., 63 Tab.,
Hardcover, ISBN 3-540-65859-9

P. J. J. Welfens
EU Eastern Enlargement and the Russian Transformation Crisis
1999. X, 151 pages. 12 Figs., 25 Tab.,
Hardcover, ISBN 3-540-65862-9

P. J. J. Welfens
Globalization of the Economy, Unemployment and Innovation
1999. VI, 255 pages. 11 Figs., 31 Tab.,
Hardcover, ISBN 3-540-65250-7

P. J. J. Welfens, J. T. Addison, D. B. Audretsch, T. Gries, H. Grupp
Globalization, Economic Growth and Innovation Dynamics
1999. X, 160 pages. 15 Figs., 15 Tab.,
Hardcover, ISBN 3-540-65858-0

R. Tilly, P. J. J. Welfens (Eds.)
Economic Globalization, International Organizations and Crisis Management
Contemporary and Historical Perspectives on Growth, Impact and Evolution of Major Organizations in an Interdependent World
2000. XII, 408 pages. 11 Figs., 20 Tab.,
Hardcover, ISBN 3-540-65863-7

P.J.J. Welfens, E. Gavrilenkov (Eds.)
Restructuring, Stabilizing and Modernizing the New Russia
Economic and Institutional Issues
2000. XIV, 516 pages. 82 Figs., 70 Tab., Hardcover, ISBN 3-540-67429-2

P.J.J. Welfens
European Monetary Union and Exchange Rate Dynamics
New Approaches and Applications to the Euro
2001. X, 159 pages. 26 Figs., 12 Tab., Hardcover, ISBN 3-540-67914-6

P.J.J. Welfens
Stabilizing and Integrating the Balkans
Economic Analysis of the Stability Pact, EU Reforms and International Organizations
2001. XII, 171 pages. 6 Figs., 18 Tab., Hardcover, ISBN 3-540-41775-3

P.J.J. Welfens, B. Meyer, W. Pfaffenberger, P. Jasinski, A. Jungmittag
Energy Policies in the European Union Germany's Ecological Tax Reform
2001. VII, 143 pages. 21 Figs., 41 Tab., Hardcover, ISBN 3-540-41652-8

P.J.J. Welfens (Ed.)
Internationalization of the Economy and Environmental Policy Options
2001. XIV, 442 pages. 57 Figs., 61 Tab., Hardcover, ISBN 3-540-42174-2

P.J.J. Welfens
Interneteconomics.net
Macroeconomics, Deregulation, and Innovation
2002. VIII, 215 pages. 34 Figs., 30 Tab., Hardcover, ISBN 3-540-43337-6

D.B. Audretsch, P.J.J. Welfens (Eds.)
The New Economy and Economic Growth in Europe and the USA
2002, XII, 350 pages. 28 Figs., 59 Tab., Hardcover, ISBN 3-540-43179-9

C.E. Barfield, G. Heiduk, P.J.J. Welfens (Eds.)
Internet, Economic Growth and Globalization
Perspectives on the Digital Economy in Europe, Japan and the U.S.
2003. XII, 388 pages. 34 Figs., 67 Tab., Hardcover, ISBN 3-540-00286-3

J.T. Addison, P.J.J. Welfens (Eds.)
Labor Markets and Social Security
Issues and Policy Options in the U.S. and Europe, 2nd edn.
2003. X, 402 pages. 56 Figs., 52 Tab., Hardcover, ISBN 3-540-44004-6

T. Lane, N. Oding, P.J.J. Welfens (Eds.)
Real and Financial Economic Dynamics in Russia and Eastern Europe
2003. XII, 293 pages. 50 Figs., 63 Tab., Hardcover, ISBN 3-540-00910-8

J.E. Gavrilenkov, P.J.J. Welfens, R. Wiegert (Eds.)
Economic Opening Up and Growth in Russia
2004. IX, 298 pages. 61 Figs., 70 Tab., Hardcover, ISBN 3-540-20459-8

P.J.J. Welfens, A. Wziątek-Kubiak (Eds.)
Structural Change and Exchange Rate Dynamics
The Economics of the EU Eastern Enlargement
2005. VI, 288 pages. 26 Figs., 58 Tab. Hardcover, ISBN 3-540-27687-4

E.M. Graham, N. Oding, P.J.J. Welfens (Eds.)
Internationalization and Economic Policy Reforms in Transition Countries
2005. XI, 340 pages. 77 Figs., 28 Tab. Hardcover, ISBN 3-540-24040-3

P.J.J. Welfens, F. Knipping, S. Chirathivat, C. Ryan (Eds.)
Integration in Asia and Europe
Historical Dynamics, Political Issues and Economic Perspectives
2006. VI, 284 pages. 32 Figs., 30 Tab. Hardcover, ISBN 3-540-28729-9

P.J.J. Welfens, A. Wziątek-Kubiak (Eds.)
Structural Change and Exchange Rate Dynamics
2005. VI, 288 pages. 36 Figs., 58 Tab. Hardcover, ISBN 3-540-27687-4

H.G. Broadman, T. Paas, P.J.J. Welfens (Eds.)
Economic Liberalization and Integration Policy
2006. VI, 358 pages. 93 Figs., 52 Tab. Hardcover, ISBN 3-540-24183-3